Maria Luiza Lopes de Oliveira Santos

Using Mineral Waste to Formulate Mortars

Maria Luiza Lopes de Oliveira Santos

Using Mineral Waste to Formulate Mortars

For the Construction Industry

Imprint
Any brand names and product names mentioned in this book are subject to trademark, brand or patent protection and are trademarks or registered trademarks of their respective holders. The use of brand names, product names, common names, trade names, product descriptions etc. even without a particular marking in this work is in no way to be construed to mean that such names may be regarded as unrestricted in respect of trademark and brand protection legislation and could thus be used by anyone.

Cover image: www.ingimage.com

This book is a translation from the original published under ISBN 978-613-9-65652-3.

Publisher:
Sciencia Scripts
is a trademark of
Dodo Books Indian Ocean Ltd. and OmniScriptum S.R.L publishing group

120 High Road, East Finchley, London, N2 9ED, United Kingdom
Str. Armeneasca 28/1, office 1, Chisinau MD-2012, Republic of Moldova, Europe
Printed at: see last page
ISBN: 978-620-7-79420-1

SUMMARY

ABOUT THE AUTHOR

Maria Luiza Lopes de Oliveira Santos

She has a degree in Civil Engineering from Universidade Potiguar, a master's degree in Civil Engineering from Universidade Federal de Santa Catarina, a PhD in Materials Science and Engineering from Universidade Federal do Rio Grande do Norte and a post-doctorate in Materials Technology from the same institution. She is currently an adjunct professor, level IV, at the Federal University of Maranhao, teaching on the Civil Engineering course.

PRESENTATION

In all industrial sectors, companies are investing in equipment and training professionals to eliminate the culture of waste and consolidate the culture of reducing losses, reusing and recycling waste. According to Moura (2000) apud Silva (2006), waste not only has an environmental impact, but its reduction and reuse can also minimize the consumption of natural and energy resources, which often implies lower economic expenditure.

According to Sjostrom (1996) apud John (2000), construction consumes between 14% and 50% of the natural resources extracted from the earth. In view of this, it is essential that the construction industry starts to consume more and more waste instead of non-renewable raw materials.

An example of consumption by the construction industry is quarrying. Crushing rocks (basalt, limestone, granite, gneiss, etc.) basically produces five types of product: quarry fines, crushed stone 0, crushed stone 1, crushed stone 2 and crushed stone 3. Of these products, the only ones that find noble applications in civil construction are gravels 1, 2 and 3, while gravel 0 and quarry fines are only used in applications without much commercial value (SAMPAIO and ALMEIDA, 2002). As for quarry fines, when they don't have a defined destination, they are stored in open piles in the quarry yard and end up causing serious environmental damage, such as silting up rivers and water courses, visual pollution and atmospheric pollution (SA, 2006).

As for the ceramics industry, it is worth noting that it produces an average of 32 tons of porcelain tile polishing waste per month. To date, there are no records of this waste being used by the construction industry (SOUZA, 2007). The waste is stored in the open air in the yard of the industry for an indefinite period, exposed to the weather and potentially causing environmental damage such as atmospheric pollution, visual pollution, among others.

According to Santin (1996) apud Silva (2006), in Brazil there are frequent building processes that use masonry walls coated with mortar, both in the internal and external sealing systems, in addition to the growing use of structural masonry. This is especially true of middle- and low-income housing, where cement-based conglomerates and mineral aggregates make up 20% of the average cost of construction work. Also showing the high consumption of mortar, data from Capozzi (1996) apud Silva (2006) indicates that 37% of the buildings under construction in the city of Sao Paulo/SP use mortar coatings with a paint finish on the façades.

The consumption of mortar in Brazilian construction works amounts to 0.145 m^3 of mortar per m^2 of construction, of which 0.030 m3 is for masonry, 0.040 m3 for subfloors, 0.030 m3 for external cladding and 0.045 m3 for internal cladding (SOUZA and FRANCO, 1997). The same author also points out that cement consumption per m2 of construction is approximately 26.05 kg/m2, where

4.32 kg is consumed in masonry, 10 kg is consumed in subflooring, 5.25 kg is consumed in external cladding and 6.48 kg is consumed in internal cladding.

Partially replacing cement with a percentage of quarry fines and porcelain polishing residue in the construction industry has several advantages, among which the following stand out:

• Full use of quarries;

• Total recycling of the porcelain polishing residue generated by the ceramics industry, since it is generally disposed of in landfills, generating costs for manufacturers;

• Reducing the consumption of cement, the production of which has major environmental impacts;

• Unquestionable solutions to environmental problems.

In addition, it is assumed that, as they are mineral waste, limestone quarry waste and porcelain polishing waste can develop pozzolanic activity.

Recycling is of fundamental importance for reducing environmental impact and has proved to be an interesting alternative for the use of waste by the construction industry, developing materials that meet technical standards, improving their properties and not posing risks to the environment (SILVA, 2005).

According to Silva and Campiteli (2006), it is possible to reduce the production costs of cement and sand mortars, as well as cement, lime and sand mortars, without compromising their performance, by replacing cement and/or lime with pulverized artificial sand, using appropriate proportions.

In relation to the porcelain polishing residue, Silva (2005) concluded that adding 10% of this residue to concrete produced satisfactory results in terms of mechanical properties and durability, thus developing a material with better performance and contributing to environmental preservation.

It should be pointed out that the contents of this book are the result of the author's doctoral thesis, which was carried out between 2004 and 2008, under the supervision and co-supervision, respectively, of Professors Antonio Eduardo Martinelli and Marcus Antonio de Freitas Melo. As such, it is possible to observe throughout the text the obsoleteness of some acronyms and technical standards cited or used in the development of the experimental program. As the work was done at a different time, it is possible that much has evolved in this period, both in terms of the corpus of analysis and the accessory technologies. Perhaps even in terms of the academic argumentation inherent in the subject. However, it is interesting to point out that the analysis in question sheds light on certain elements in a way that has contributed and may yet contribute to the development of new analysis/work inherent to the topic addressed in this theme.

Book structure

The structure of this book is made up of six chapters, the first of which sets out the importance of the topic, the justifications and the objectives of the research.

The second chapter presents the state of the art on the main mineral residues used in the manufacture of cementitious matrix materials.

The third chapter deals with the theoretical framework on cement, mortars, limestone quarry waste and porcelain polishing waste.

The fourth chapter contains the experimental program and the tests carried out.

Then, in the fifth chapter, the results of the research are presented, analyzed and discussed.

The sixth chapter presents the final considerations.

After the sixth chapter are the references and annexes.

Maria Luiza Lopes de Oliveira Santos

CHAPTER 1

INTRODUCTION

Once a subject limited to scholars and activists, the issue of the environment has become a global issue and is now present in the most diverse segments of society. Concern about preserving the environment is already seen as a priority by the public authorities and society's support is becoming increasingly necessary (OLIVEIRA, 2008).

In view of this, modern engineering is obliged to work with three variables that are considered basic and have historically been ignored: environmental responsibility, social responsibility and sustainability. The fact is that engineering must get closer to real human needs, especially since the construction industry is a key player in the current environmental pollution scenario (DEL CARLO, 2008).

Sustainable development must meet all the needs of present generations without compromising the needs of future generations. Today, the interdependence between the concepts of the environment, waste recovery, sanitation and health is quite clear, but action in these sectors must be integrated and aimed at improving society's quality of life.

Schenini et al (2004) point out that the state plays an extremely important role in the context of sustainability. As an example of the role of the state, we can cite the creation and approval of laws that encourage and induce society towards sustainable development, as well as the implementation of public policies aimed at developing a culture of preservation.

Everything that surrounds us and that we use in our daily routine will one day be waste: houses, toothbrushes, cars, furniture, bridges, airplanes, televisions, mattresses, clothes, shoes and so on. To this total, we must add all the waste from the process of extracting raw materials and producing goods. Thus, in any society, the amount of waste generated exceeds the amount of goods consumed. Industrial society, by multiplying the production of goods, has aggravated this process (ROCHA and JOHN, 2003).

Recycling, which is defined here as the reuse of materials as raw materials for a new product, has become more widespread in recent decades. The consumption of natural resources and energy has increased in proportion to the growth and development of the world's population. Environmental concerns include excessive or inefficient consumption of non-renewable natural resources (LEITE, 2001).

The development of recycling has been assiduously seeking to improve the quality of life from an environmental point of view, thus reducing the waste generated by industrial activity. In the business world, awareness is beginning to grow that the generation of waste generates costs, as it requires treatment and disposal in properly regulated and appropriate locations. As a result, various industries are looking for options to reduce costs in order to reduce environmental impacts and increase their credibility in the consumer market (SILVA, 2005).

Silva (2005) also points out that recycling has achieved significant economic, social and environmental results. Its main advantages include:

• Reducing the amount of waste sent to industrial landfills, saving transportation costs and reducing degraded areas;

• Saving various natural resources;

• Improving the company's image in the eyes of its customers;

• Reducing industrial costs by incorporating waste into various production processes.

Recycling is undoubtedly one of the best alternatives for reducing the impact that the environment can suffer from the consumption of raw materials, whether renewable or not, and the disorderly generation of waste. In recent years, waste recycling has been encouraged worldwide, whether for political, economic or ecological reasons (LEITE, 2001).

The construction industry is considered to be the biggest consumer of natural raw materials, thus contributing significantly to further environmental degradation. There is also an indirect contribution, such as the cement industry which, according to Capello (2008), accounts for 5% of the world's carbon dioxide emissions. This is because the production process for each ton of clinker (its main component) releases the same amount of CO_2 into the atmosphere.

Cement is the basic input of the construction industry, used in large quantities in the production of housing, infrastructure, roads, bridges, sanitation and hydroelectric dams. The cost of cement in relation to the total cost of a residential building project is around 7% to 9% (TEIXEIRA, 2008). According to a report published on the Cimento.org website (2008), sales to the domestic market in the period from April 2007 to March 2008 reached 46.1 million tons of product.

In the last decade in particular, the construction industry has been seeking to use materials that are durable, and this characteristic must be evaluated with the same attention as mechanical properties and initial cost. As a result, research has been carried out into resistant materials capable of satisfactorily meeting the performance and important and desirable characteristics of each material. The use of composite materials has therefore become a viable alternative. According to Budinsky (1996), composite materials can be defined as a mixture of two or more different materials with characteristics inferior to those of the resulting material.

The use of mineral residues has proven to be very important in the performance of materials with cementitious matrices, both in their fresh and hardened properties, as well as in terms of durability. Incorporating waste into the production of these materials can save energy and contribute to sustainable development (SILVA, 2005). The volume of industrial by-products, with or without pozzolanic properties, produced worldwide is much greater than the demand for recycling. Therefore, the study of replacing part of the cement in concrete or mortar with waste with pozzolanic or filling properties is of great importance.

An industrial by-product is the fine material from the crushing of quarries and the comminution of rocks. In Brazil, this by-product has various names, such as stone dust, industrial sand, artificial sand (LANG and ROCHA, 2006) or quarry fines. In this research, these fines will be called limestone quarry residue (RPPC). This material is still being used discreetly in Brazil. Reinforcing the idea of sustainability, and with the aim of reducing the environmental impacts generated by quarry waste as much as possible, we decided to use it in this research in the form in which it is stored in the quarry yard.

In addition to the use of limestone quarry residue (RPPC), and in line with the search for the use of waste, with the aim of greater environmental preservation, porcelain polishing residue was also used, in this research called RPP, considering the great environmental impact generated by the ceramic tile industry.

According to Silva (2005), the ceramics industry for internal and external cladding used in construction produces a large amount of waste. Some of it is reused in the sector itself and some is discarded. The porcelain tile manufacturing process produces a lot of waste that has not yet been used by the industry. One of the residues produced in the porcelain tile industry comes from polishing, which is necessary to finish the piece.

The general aim of this book is to show the feasibility of substituting 5%, 10%, 20% and 30% of cement with limestone quarry waste or porcelain tile polishing residue in mixed mortars for laying and coating, without damaging their performance.

In order to meet the main objective, the formulated mortars were characterized using consistency index tests, water retention, mass density in the fresh state, incorporated air content, apparent mass density in the hardened state, modulus of elasticity, flexural tensile strength and water absorption by capillarity, in order to evaluate the influence of limestone quarry waste and porcelain tile polishing waste on the properties and microstructure of the mortars.

CHAPTER 2

STATE OF THE ART

2.1 USE OF MINERAL WASTE IN CIVIL CONSTRUCTION

According to Selmo (1989) apud Silva (2006), Boltshauer's books (1963) show that in Greece, around 2000 B.C., raw clay was used in various rammed earth or wattle and daub constructions involving sturdy wooden structures. The same technique was used in Roman architecture between the 7th and 6th centuries BC. In modest Egyptian constructions, from 1600 to 1100 BC, clay paste was used to encase the structural framework of houses made from plant stems or palm trunks. Clay baked into molded slabs with low reliefs was also used in Greek architecture to cover the stone walls of facades. The Greeks and Romans prepared mixtures of lime, sand and water, or added plaster to obtain stucco, which is a specialty for interior cladding. In some cases, they also added marble dust to the mixture. The Romans made extensive use of mortars, both for laying masonry and for cladding their buildings. For over 2000 years, mortar has been used to lay and coat the stone blocks that make up the walls of buildings.

According to Westphal (2004) apud Silva (2006), mortar began to be used in Brazil in the first century to lay stone masonry (widely used at the time). The lime used in mortar was obtained by burning shells and shellfish. Whale oil was also widely used as a binder at the time.

The recycling of industrial waste for use as alternative raw materials is not new and has been successfully carried out in several countries. The construction industry is a branch of technological activity which, due to the volume of natural resources consumed, seems to be very suitable for absorbing solid waste (AHMED, 1996 apud COIMBRA et al, 2004).

Since ancient times, man has depended on rocks, first used as shelter and then as weapons and tools. Today, even with all the sophistication and technology available to us, our dependence on mineral resources is indisputable (AMORIM et al, 1999).

Mineral addition to cement is not a recent process, originally incorporating natural pozzolans into the mix and some time later, industrial ash, with properties similar to the former. Pioneering research with bottom ash dates back to the 1930s in the United States (GARCIA et al, 2006).

Mineral goods are of significant importance to society, and no civilization can do without their use, especially when it comes to quality of life, since the basic needs of human beings are essentially met by these resources (KUMMER et al, 2007).

Research is moving towards the use of inert additions to cement, aimed not only at chemical and mechanical properties, but also at reducing environmental problems caused by the generation of tons of mineral waste (GARCIA et al, 2006).

Below is a state-of-the-art overview of the main mineral residues used in the manufacture of cementitious matrix materials. It should be noted that waste from rock quarrying, known as quarry fines, and from polishing porcelain tiles, both used in this research, will be covered in the third chapter of this book.

2.2 cementitious materials with mineral waste

2.2.1 Foundry sand

The metal industry has historically been known as a sector that produces considerable volumes of solid waste inert to its production processes, among which foundry sand stands out. Despite the implementation of new technologies and greater stringency on the part of environmental agencies, large volumes of solid waste continue to be generated and still cause concern for entrepreneurs in the sector, mainly due to environmental liability laws and the relatively high cost of managing and disposing of this waste (CRUZ, 2002).

Among the waste produced by the metal industry, foundry sand is the most widely used, as it has similar characteristics to the fine aggregate commonly used in the construction sector (PEREIRA, 2002 apud NASCIMENTO et al, 2007). Thus, various studies have been carried out on the use of this waste as aggregate, with the aim of reusing it and thus generating an environmentally correct final destination for it (NASCIMENTO et al, 2007).

Foundry sand is made up of base sand, where silica in the form of quartz is the main component and comes from natural reserves (MARIOTTO, 2001 apud NASCIMENTO et al, 2007); bentonite and coal dust are considered to be foundry sand additives. Base sand is considered an inert material within the cement paste (WATANABE, 2004 apud NASCIMENTO et al, 2007).

Studies have shown that the incorporation of bentonite in pastes promotes the formation of ettringite and increases the Si/Ca ratio, a factor responsible for the formation of C-S-H, which demonstrates an increase in the degree of crystallinity of the material. On the other hand, pastes made with coal dust do not have a significant influence on cement hydration; it only favors the formation of ettringite. Analysis using scanning electron microscopy shows that an increase in the number of additions leads to a greater formation of ettringite crystals, especially in samples containing bentonite (NASCIMENTO et al, 2007).

2.2.2 Battery waste

Resolution 257/99 published by CONAMA (National Environment Council) has outlined the fate of batteries in the country. Article 6 of the law states that, as of January 1, 1999, the manufacture, import and sale of batteries must comply with the following limits for active metals in batteries: 0.010% mercury by weight, 0.015% cadmium and 0.200% lead (COIMBRA et al, 2004).

In this context, we find zinc-carbon batteries, which are widely used due to their ease of use and reasonable performance or the ratio (battery characteristic)/price. The use of this type of battery has increased significantly in recent decades, due to the fact that it does not contain heavy metals such as mercury and introduced the concept of "heavy metal-free batteries". Currently, the battery industry uses around 4% of the zinc produced in Brazil, which is equivalent to approximately 8,000 tons a year, which are used and then discarded in household waste (COIMBRA et al, 2004).

According to Coimbra et al (2004), concrete with an addition of 1% by weight of zinc-coal pile increases mechanical resistance to compression and reduces deformation compared to concrete without an addition. However, with 10% addition, the concrete did not have enough agglomerating phase to be resistant. It even showed characteristics of "crumbling" and loss of sand and gravel particles, not maintaining homogeneity in its use.

In the composition containing 1% addition, the formation of needles occurred (Figure 1), which allowed the phases of the pile to intertwine with the cement paste, which most likely led to an increase in mechanical strength, *while in the* composition with 10% addition, there may have been interference by the constituents of the pile in the hydration reactions of the cement paste, as seen experimentally by the typical smell of ammonia in the mixture (COIMBRA et al, 2004).

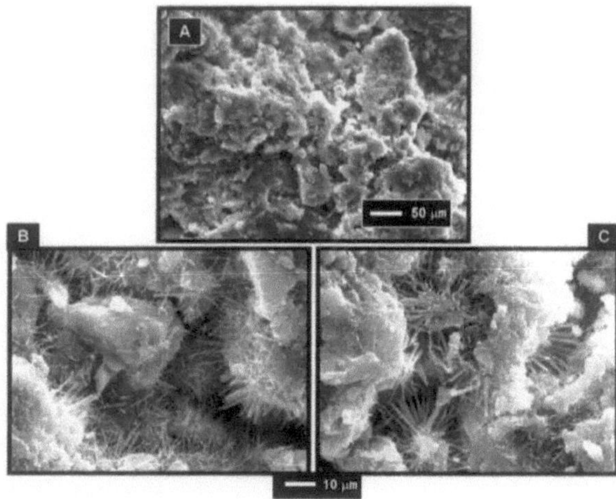

Source: COIMBRA et al, 2004.

Figure 1 - Micrographs of concrete modified with 1% by weight of zinc-carbon pile, after 28 days of curing at 23 °C and humidity above 95%. A) General aspect of the sample (1,000x magnification); B) and C)

Details of the needles (5,000x magnification).

According to Coimbra et al (2004), concretes with additions have larger pore sizes and total open porosity than those without additions, which could lead to a reduction in mechanical strength values. However, this is probably due to the combination of the phases of the batteries in the concrete, in the case of the composition with 1% addition, with the formation of the hydrated phases of the cement, thus making a mutual contribution to the improvement in their mechanical compressive strength.

2.2.3 Marble and granite cutting waste

Concern about the environment is growing sharply all over the world. Of course, the marble and granite industry is a constant cause of environmental impact (CALMON et al, 1997).

Brazil has a large reserve of ornamental stones (granite and marble) with the most varied aesthetic aspects. Natural ornamental stones are the main competitors for ceramic tiles (MOREIRA et al, 2003).

The system of splitting ornamental stone blocks to produce slabs, using equipment called looms, generates a significant amount of waste in the form of sludge (Figures 2 and 3). This slurry is generally made up of: water, grit, lime and ground rock, as well as metal fragments from the wear of the blades. Lime can be replaced by other materials, such as: ground limestone, blast furnace slag, among others (CALMON et al, 1997).

Source: PONTES and VIDAL, 2005.

Figure 2 - Discharge of mud into the tailings dam.

Figure 3 - Sampling in the slurry discharge pipe at the tailings dam.

According to Mothe Filho et al (2001) apud Moreira et al (2003), sludge when dry becomes a non-biodegradable solid waste classified as class III - inert. However, when this waste is not disposed of correctly, it can reach rivers, ponds, streams and even natural water reservoirs, causing siltation problems.

According to Moura et al (2002), around 25% to 30% of the block is transformed into powder during sawing, which is deposited in company yards. In Brazil, the estimated joint generation of marble and granite cutting waste is 240,000 tons/year.

According to Freire and Motta (1995) apud Calmon et al (1997), some of the residue's specific characteristics are favorable to its industrial use, such as its fine granulometry, its predefined chemical composition and the lack of mixed grains among the basic components; in addition, this residue is potentially applicable as mortar, raw material for ceramics and mineral fillers.

The use of marble and granite cutting waste in civil construction aims to contribute directly to reducing the environmental impact resulting from the high consumption of natural aggregates (MOURA et al, 2002).

Studies carried out on the use of marble and granite cutting residue (RCMG) as a substitute for part of the sand in the production of mortars for cladding show that this residue is in crystalline form and does not present any environmental risks. In terms of compressive strength, mortars with RCMG performed better than mortars with no additions (MOURA et al, 2002).

According to Calmon et al (1997), when lime is replaced by RCMG, it can be concluded that:

• The mortars retained more water and exuded more water, but these phenomena occurred discreetly. Although these phenomena affect the workability of the mortars, it is necessary to evaluate their adhesion behavior;

• The mortars obtained substantial improvements in axial compressive strength, diametrical compressive tensile strength, water absorption by capillarity, absorption by immersion, void ratio and specific mass;

•   The residue may be working as a filer.

2.2.4 Construction and demolition waste

In simpler terms, this material can be defined as waste from the construction, repair, renovation and demolition of structures and roads (LEITE, 2001).

In any country in the world, the construction industry is the sector responsible for consuming the largest volume of non-renewable natural resources, it is estimated that the resources extracted vary between 15% and 50%; in addition, its products are major energy consumers (JOHN, 1999 apud MIRANDA, 2000).

Construction and demolition waste (CDW) accounts for 50% of the mass of municipal solid waste (MSW). One estimate puts the amount at 68.5 million tons per year, given that 137 million people live in urban areas. Most countries in the world invest in a formal management system to reduce illegal and systematic dumping, which causes silting of rivers, clogging of culverts, degradation of areas and depletion of landfill areas, as well as high socio-economic costs, especially in medium-sized and large cities (ÂNGULO, 2005). In Brazil, the per capita generation of CDW has been estimated at 500 kg/inhabitant per year, the median for some Brazilian cities (PINTO, 1999).

Countries like the Netherlands, Denmark and Germany have CDW recycling rates of between 50% and 90%, while countries like Portugal and Spain have rates of less than 50%. In Portugal, the small scale of production at recycling plants for the mineral fraction of CDW, most of which belongs to the public sector, means that recycling rates are modest (ÂNGULO, 2005).

Traditionally, the majority of CDW is sent to landfills, often the same ones used for municipal solid waste (MSW), which do not always offer the conditions that would be desirable for dealing with the various residues they receive (PEREIRA et al, 2004).

In Brazil, as in other countries, the recycling of the mineral fraction of CDW generates aggregates for paving and filling material for landfills. It is used less in the manufacture of cement-based products (concrete, blocks, mortar, etc.) (ÂNGULO, 2005).

According to Amorim et al (1999), mortars with CDW have satisfactory mechanical performance, making the recycling of waste very promising and capable of producing low-cost construction elements.

When finely ground, CDW can exhibit pozzolanic activity and be used with lime or as a partial substitute for Portland cement in the production of grout, mortar and concrete (BARONIO and

BINDA, 1997 apud FARIAS FILHO et al, 2006).

Research carried out by Levy (2001) confirms that the addition of concrete and masonry waste up to 20% does not affect the behavior of the concrete compared to the reference (without waste), demonstrating that it can be used without any restrictions in terms of strength and durability.

According to research carried out by Leite (2001), the absorption rates of recycled aggregates are quite high and, therefore, there is a need to compensate for them during the production of concrete so that there is no excessive reduction in the free water content of concrete mixtures produced with recycled aggregates, which would compromise workability. Another relevant factor is that the tensile strength of concrete with recycled aggregates is inversely proportional to the size of the recycled aggregate used, due to the inherent fragility of the material.

2.2.5 Granulated blast furnace slag

According to Pecchio and Battagin (1999) apud Brizola (2007), blast furnace slag is a by-product of the manufacture of pig iron in blast furnaces, each ton of which produces approximately 300 kg of slag. Slag is the supernatant liquid at the base of the blast furnace and is separated from the pig iron, which is also liquid, by a difference in density. At the exit of the blast furnace, the slag undergoes rapid cooling by falling into water tanks, known as granulation tanks. This is the process of obtaining the granulated vitreous slag that is traditionally used by the cement industry as an addition to Portland cement.

According to Smolczyk (1980) apud Melo Neto (2002), sudden cooling prevents the organization of a crystalline structure (Figure 4), retaining energy in the atomic structure of the vitreous phase, thus characterizing the material as amorphous. When the slag is cooled slowly, it is largely crystalline and has almost no agglomerating properties.

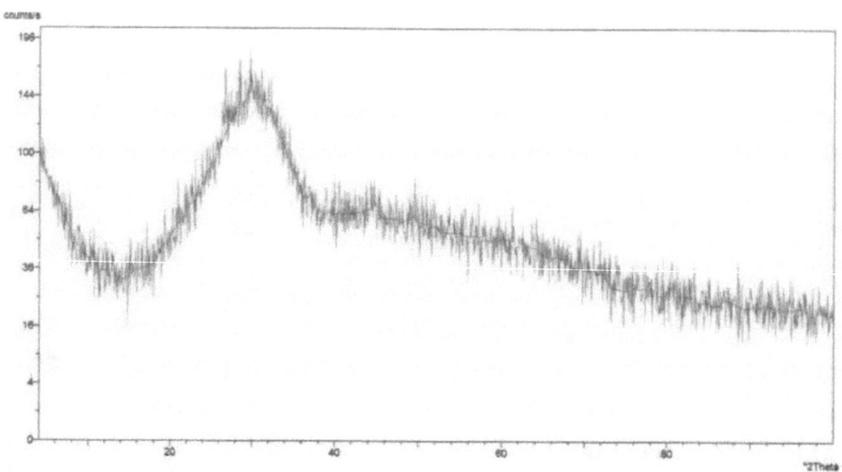

Source: VOGT et al, 2006.

Figure 4 - X-ray diffraction of the ground slag sample.

Cincotto et al (1992) apud Brizola (2007), state that the percentages of the main components of Brazilian blast furnace slag are: 30% to 35% silica, 40% to 45% calcium oxide, 11% to 18% aluminum oxide, 2.5% to 9% magnesium oxide and 0% to 2% Fe_2O_3.

Although its fundamental hydration mechanism is the same as that of cement, unlike cement, when slag comes into contact with water it does not hydrate or does so very slowly (ILLER, 1979 apud BRIZOLA, 2007).

However, there are activation methods that are used to accelerate hydration. Breugel and Rooij (2004) apud Brizola (2007) report that slag is most commonly activated by the use of Portland cement, which requires between 10% and 20% for activation to occur.

Blast furnace slag is a waste product commonly used as an addition to concrete (GOLASZEWSKI, 2005), as it has good cementitious properties due to its reactive properties (RESCHKE et al, 2005), affecting the evolution of the compound's strength over time and its durability (VOGT et al, 2006).

The presence of blast furnace slag makes the microstructure of hydrated cement paste denser, due to the greater formation of C-S-H, which improves mechanical strength and long-term durability (NEVILLE, 1997). The greater density of the matrix provided by blast furnace slag increases the concrete's resistance to chloride penetration and sulphate attack (MOURA, 2000).

According to research carried out by Golaszewski (2005), the use of blast furnace slag significantly improves the performance of concrete, particularly with regard to durability. The addition of blast furnace slag significantly influences the rheological properties, setting time and compressive strength of mortars. The form and magnitude of this influence depends on the properties of the cement and superplasticizer used in the dosages, as well as the properties and content of the blast furnace slag added.

The manufacture of concrete by partially replacing the cement of the reference concrete by mass with blast furnace slag showed that the 35% addition promotes a slower increase in strength, but does not significantly affect the compressive strength value at 28 days. In addition, it contributes to a 10% increase in the dynamic modulus of elasticity at this age. Fracture toughness does not seem to be affected by the additions (VOGT et al, 2006).

2.2.6 Steering wheel gray

The use of mineral additions began with the use of fly ash in mass concrete dams, with the aim of reducing the heat of hydration and the cost of concrete (LANA, 2005).

Fly ash is the mineral coal residue used in thermal power plants (LANA, 2005). This waste is considered a pozzolan and is one of the artificial ones, i.e. it is a secondary product of a process (by-product), potentially polluting the environment, but with an amorphous (vitreous) structure due to the high temperatures at which it is generated (SILVEIRA, 2004). X-ray diffraction analysis shows that the main constituents of fly ash are quartz and mullite (VAGHETTI, 1999).

There are strong arguments in favor of using this ash in concrete, especially in the South, where there is an abundance of this material. Rio Grande do Sul holds approximately 90% of the country's coal reserves (around 28 billion tons of coal) - source: IRGA, 97/98 harvest cited by Vaghetti (1999).

The Candiota Thermoelectric Plant produces around 1,000,000 tons of ash every year, 80% of which is fly ash and 20% is heavy ash (SILVA et al, 1999 apud LANA, 2005).

The 800,000 tons of fly ash have the characteristics to be used as a mineral addition in concrete. In addition, its disposal in the environment can cause serious pollution problems (LANA, 2005).

When fly ash has adequate fineness, its reactivity is good and it has a considerable influence on the workability and speed of development of the strength of hardened concrete (VAGHETTI, 1999).

Uchikawa (1986) apud Silveira (2004) points out that fly ash slows down the hydration of C_3S (tricalcium silicate) at early ages and speeds up the hydration of this compound at later ages. In view of this, Lana (2005) states that concrete made with fly ash is more sensitive to moisture content at early ages, thus requiring longer curing.

The greatest benefit of incorporating fly ash into concrete is that it makes fresh concrete more workable, making it easier to cast and finish. It also causes better cohesion and less tendency to ooze than concrete made with cement alone (FRIZZO, 2001 apud PETRY, 2004). This material also has a beneficial effect on reducing porosity and increasing concrete strength in the long term (SILVEIRA, 2004).

According to Neville (1997), the incorporation of supplementary materials, especially fly ash, into Portland cement concrete can increase its properties in both the fresh and hardened states, thus providing greater durability.

CHAPTER 3

THEORETICAL FRAMEWORK

3.1 CLAY

3.1.1 Definition

Mortars are construction materials that are being used more and more frequently and are more specialized in buildings. Their traits and characteristics depend on the type of application (RODRIGUÉZ, 1994).

NBR 13281 (ABNT, 2005) defines mortar as a homogeneous mixture of fine aggregate(s), inorganic binder(s) and water, with or without additives, with adherence and hardening properties, which can be dosed on site or in its own installation (industrialized mortar).

3.1.2 Types of mortar

NBR 13281 (ABNT, 2005) classifies and defines the various types of mortar according to their use and application:

a)   Laying mortar

•   Mortar for sealing masonry: suitable for bonding sealing components (such as blocks and bricks) when laying masonry, with a sealing function;

•   Mortar for laying structural masonry: suitable for bonding sealing components (such as blocks and bricks) when laying structural masonry;

•   Mortar to complete the masonry (wedging): suitable for sealing after the last row of components.

b)   Mortar for coating walls and ceilings

•   Mortar for internal cladding: suitable for cladding internal areas of the building, characterized as a smoothing layer (plaster or single layer);

•   Mortar for external cladding: suitable for cladding façades, walls and other building elements in contact with the external environment, characterized as a smoothing layer (plaster or single layer).

c)   General purpose mortar

Suitable for laying non-structural masonry and cladding internal and external walls and ceilings.

d)   Plastering mortar

Suitable for covering plaster, providing a fine surface to receive the finish; also called fine putty.

e)   Thin-layer decorative mortar

Finishing mortar suitable for decorative coatings, in a thin layer.

f)   Decorative single-layer mortar

Finishing mortar suitable for coating façades, walls and other building elements in contact with the outside environment, applied in a single layer for decorative purposes.

3.1.3 Mixed mortar constituent materials

3.1.3.1 Cement

3.1.3.1.1 Definition

Cement, in the general meaning of the word, can be considered to be any material with adhesive and cohesive properties capable of binding mineral fragments together to form a compact whole. This definition covers a wide variety of materials. In the construction field, the meaning of the term is restricted to binding materials used with stones, sand, bricks and blocks, among others (NEVILLE, 1997).

3.1.3.1.2 History

The use of cement goes back a long way. The ancient Egyptians used impure calcined gypsum. The Greeks and Romans used calcined limestone and later learned to mix lime and water, sand and fragmented stone, bricks or shattered tiles. It was the first concrete in history (NEVILLE, 1997).

In 1756, engineer John Smeaton, in charge of building the Eddystone lighthouse off the Corn coast in England, was looking for a binder that would harden even in the presence of water, in order to make the job easier. In his attempts, he found that a calcined mixture of limestone and clay became, once dry, as resistant as the stones used in construction. In 1824, however, it was the bricklayer Joseph Aspdin who patented the discovery, naming it Portland cement, in reference to Portlandstone, a type of sandstone widely used in construction in the Portland region of England. The patent application stated that the limestone was ground with clay in a wet environment until it turned into powder. The water was evaporated by exposure to the sun or by heat irradiation through a steam pipe. Blocks of the dry mixture were calcined in ovens and then ground very fine (TAYLOR, 1967 apud SILVA, 2006).

3.1.3.1.3 Portland cement production

The process of making Portland cement essentially consists of grinding the raw material, mixing it intimately in the appropriate proportions and burning this mixture in a large rotary kiln up to a temperature of around 1450 °C. At this temperature, the material undergoes incipient fusion to form pellets, known as clinker. The clinker is cooled and ground to a very fine powder with the addition of a little gypsum, resulting in the commercial Portland cement widely used throughout the world (NEVILLE, 1997).

The raw materials can be mixed and ground using two types of process: wet or dry. In the wet process, the grinding and homogenization of the raw material mixture is carried out in the form of a slurry containing 30% to 40% water. Modern cement plants prioritize the dry process, which is more energy-efficient than the wet process because the water used to produce the slurry must be evaporated before the clinkerization operation. For this operation, the kilns used in the dry process are equipped with pre-suspension heaters, which allow for an efficient exchange of heat between the hot gases and the raw material mixtures, requiring a fossil fuel energy consumption of around 800 kcal/kg of clinker compared to around 1400 kcal/kg for the wet process kilns (MEHTA and MONTEIRO, 1994).

3.1.3.1.4 Chemical composition of cement

The raw materials used to make Portland cement consist mainly of limestone, silica, alumina and iron oxide. These compounds interact with each other in the kiln to form a series of more complex products and, with the exception of a small residue of lime that has not had enough time to react, a state of chemical equilibrium is reached. However, this equilibrium is not maintained during cooling, and the speed of cooling influences the degree of crystallization and the amount of amorphous material present in the cold clinker. The properties of this amorphous material, known as the vitreous phase, differ considerably from those of crystalline compounds with a nominally similar chemical composition. Another complication arises from the interaction between the liquid part of the clinker and the crystalline compounds already present (NEVILLE, 1997).

The main compounds in cement, the proportions of which are obtained through chemical analysis, are: tricalcium silicate (C_3 S), dicalcium silicate (C_2 S), tricalcium aluminate (C_3 A), tetracalcium ferroaluminate (C_4 AF) and gypsum (C H_{52}). In addition to these, which are the main constituents of cement, minor constituents are also present, such as alkalis (Na_2O; K_2O), which are present in the form of sulphates, magnesium oxide (MgO), free lime (CaO) and others (SOUZA and RIPPER, 1998).

3.1.3.1.5 Types of Portland cement

In Brazil, as in most countries with significant production, various types of cement are manufactured, which correspond to the greater or lesser enhancement of the properties of its main components and result in the development of varied mechanical, physical and chemical characteristics. Choosing the most suitable type of cement for a particular job or service requires a good knowledge of the characteristics of the various types available on the market (SOUZA and RIPPER, 1998). Table 1 summarizes the various types of Brazilian cement and the standards that specify them.

Designations	Classes	ABNT standard
Ordinary Portland cement	CP I - 25 CP I - 32 CP I - 40	NBR - 5732
Ordinary Portland cement (with additions)	CP I - S - 25 CP I - S - 32 CP I - S - 40	NBR - 5732
Composite Portland cement (with slag)	CP II - E - 25 CP II - E - 32 CP II - E - 40	NBR - 11578
Composite Portland cement (with pozzolan)	CP II - Z - 25 CP II - Z - 32 CP II - Z - 40	NBR - 11578
Composite Portland cement (with filer)	CP II - F - 25 CP II - F - 32 CP II - F - 40	NBR - 11578
Blast-furnace Portland cement	CP III - 25 CP III - 32 CP III - 40	NBR - 5735
Pozzolanic Portland cement	CP IV - 25 CP IV - 32	NBR - 5736
High early strength Portland cement	CP V - ARI	NBR - 5733

Source: SOUZA and RIPPER, 1998.

Table 1 - Types of Portland cement.

According to Roman et al (1999), cements with a higher specific surface area have the potential to make mortars more workable and with greater water retention. Slower-setting cements can produce more resilient mortars (with a greater capacity to absorb small deformations).

3.1.3.2 Cal

3.1.3.2.1 History

Man probably got to know lime at the dawn of the Stone Age. Despite this, lime only gained more prominence in archaeological documentation during and after the time of the construction of the monumental pyramids, an extraordinary achievement of the peoples of Asia Minor and, above all, of the admirable Egyptian empire. From Egypt, the art of manipulating lime moved on to Greece, then to Rome and then to other Mediterranean and surrounding regions, until it reached modern times (GUIMARÂES, 2002).

3.1.3.2.2 Types of lime

3.1.3.2.2.1 Quicklime

When limestones and dolomites are calcined (transformation of calcium carbonate - $CaCO_3$, at a high temperature, into calcium oxide - CaO) at a temperature close to 1000 °C, they produce quicklime by losing part of their constituents (carbonic anhydride - co_2) (SILVA, 2006).

When it comes from pure rock, quicklime is a white inorganic product. When it has a creamy, yellowish or slightly grayish color, it contains impurities. Its odor is equivalent to "earthy". It is born with a crystalline structure, in single crystals or crystalline conglomerates, of varying sizes and intercrystalline spaces, observable only by microscopy. Sometimes, its observation is more difficult, as its configurations appear compact or microcrystalline (GUIMARÂES, 2002).

3.1.3.2.2.2 Hydrated lime

Hydrated lime results from the continuation of the industrial process that gives rise to quicklime. The chemical reaction that produces it in the presence of water is a classic "solid-liquid phase reaction". Depending on the volume of water used, the final product can be dry or have the appearance of cream, mud, milk or a saturated solution (GUIMARÂES, 2002).

3.1.3.2.2.3 Hydraulic lime

In general, hydraulic lime can be classified as an intermediate product between virgin lime and Portland cement. It results from the calcination of clayey limestones, a procedure which produces a combination of calcium hydroxide with silica quartz and clay minerals, forming a relatively high percentage of compounds with hydraulic properties (GUIMARÂES, 2002).

3.1.3.3 Small aggregates

3.1.3.3.1 Definition

NBR 7211 (ABNT, 1983) defines fine aggregates as aggregates whose grains pass through the sieve with a mesh opening of 4.75 mm and are retained on the sieve with a mesh opening of 150 μm.

3.1.3.3.2 Function

According to Mineropar (2004) apud Silva (2006), the main functions of aggregates are:

- Providing the binder with a relatively economical filler;

• Provide the paste with particles adapted to resist applied loads, mechanical wear and weathering;

• Reduce the volume variations resulting from the setting process, hardening and humidity variations in the cement, lime and water paste.

3.1.3.3.3 Classification

According to Mineropar (2004) apud Silva (2006), aggregates are classified as follows:

• Natural: already found in nature in its final form of use (river sand);

• Artificial: those that require textural modification to reach the right condition for use, such as sand from the crushing of rocks such as basalt, limestone, among others.

3.1.3.4 Water

In addition to allowing the mortar to harden by hydrating the cement, water is also responsible for the mortar's workability. The amount of water must allow the mortar to be applied properly, but must not cause the constituents of the mortar to segregate (ROMAN et al, 1999).

3.1.4 Functions and Features

According to Fiorito (2003), cement and sand, cement, lime and sand or cement, additive and sand mortars are commonly used on construction sites and have multiple applications. The purpose of the mortar determines the type of binder or the mixture of different types of binders. Their composition and thickness when applied vary greatly.

The efficiency of a mortar depends on the quality of the materials that make it up, as well as the application of the appropriate mixes for each specific job (RIPPER, 1995). The misuse of mortars on site leads to poor performance in the short, medium and long term, resulting in continuous and costly maintenance (PINTO, 1996).

The trace of a mortar is understood to be the indication of the proportions of its components. By using the weight (mass) of the mortar, we would have absolute certainty as to the quality of the mortar and its quantities in terms of consumption and cost appropriation. However, this is impractical on the building site (FIORITO, 2003).

The mortars most commonly used in the construction industry are those used as laying mortars or coating mortars. The main functions and characteristics of each will be discussed below.

3.1.4.1 Laying mortar

According to BS 5628 (1985), when used to lay masonry walls, mortar has the function of:

• To join the masonry units together and help them resist lateral forces;

- Evenly distribute the loads acting on the wall over the entire load-bearing area of the blocks;

- Absorb the natural deformations to which the masonry is subjected;

- Seal joints against rainwater penetration.

To perform well, it must have the following characteristics (SABBATINI, 1986):

- Workability;

- Water retention capacity;

- Initial resistance;

- Adequate adhesion to the substrate;

- Durability;

- Sufficient strength to accommodate all deformations without cracking.

The choice of mortar type can also be based on the type of element to be laid, or on construction standards, or even according to the stresses to be borne by the masonry. For example, a mortar can be correlated with the type or characteristics of the elements to be laid: concrete blocks; ceramic blocks; solid bricks; perforated bricks; lightweight concrete slabs, etc. Thus, a masonry element with a high initial water absorption rate must be compatible with mortar with high water retention (FIORITO, 2003).

3.1.4.2 Coating mortar

According to Cincotto et al (1995), when used as a cladding, mortars can simply cover wall irregularities and form part of the building's aesthetic finish, but they can also contribute to safety and habitability requirements. Mortars must therefore have the following characteristics:

- Mechanical and dimensional stability (tensile, compressive, impact and abrasion resistance) and fire resistance;

- Contribute to the watertightness of the wall, forming a barrier to water penetration;

- Contribute to the hygrothermal and acoustic comfort of the environment;

- It should be applied by checking the contribution of the base to the performance of the whole;

- It must have compatibility between its constituent materials and the materials of the base on which it will be applied.

It is worth remembering that, as the layers of coatings are connected to each other, the deformation of any of them due to endogenous causes or external forces will result in stresses acting on each layer. The stresses acting on coatings are of particular interest (FIORITO, 2003).

3.1.5 Main properties

Evaluating the performance of building materials is considered to be of great importance (KRUS et al, 1997). In order to predict the performance of building materials when exposed to the elements, their properties need to be studied prior to their application. Analyzing the relationship between microstructure and properties contributes positively to their performance (QUENARD et al, 1998).

It can basically be said that mortars need the same properties in order to perform well when used. The following are the properties required of laying and coating mortars, including their characterization.

3.1.5.1 Fresh properties

3.1.5.1.1 Workability

Workability can be considered a subjective property of mortars. There is no standard workability to be used for both laying and coating masonry. RILEM MR 3 (1982) prescribes that workability is a complex property, basically characterized by the consistency and plasticity of mortars.

A mortar can be considered to have good workability when it spreads easily over the substrate and adheres to its surface, and in the case of coating mortars, when it provides ease in the final finish.

According to Roman et al (1999), good mortar workability depends on a combination of various factors such as the quality of the aggregates and binders, the amount of water used in the mix, consistency, the mortar's water retention capacity, the time taken to prepare the mortar, adhesion, fluidity and mass. Another important factor to be considered in this property is the sand's granulometric characteristics: uniformity coefficient, fineness modulus and grain shape (TRISTÂO, 1995).

Evaluating, measuring and prescribing the workability values of mortars scientifically, by means of tests, is a very difficult task, since this property depends not only on the intrinsic characteristics of the mortar, but also on the skill of the professional carrying out the work, the various properties of the substrate and the application technique (CASCUDO et al, 2005).

3.1.5.1.2 Consistency and consistency retention

Consistency is the property by which fresh mortar tends to resist deformation. The amount of water added to the mortar is the main factor influencing this property, making it the main property used to control the amount of water in mortars. Other factors such as the ratio of water to binder, the ratio of binder to aggregate, the grain size of the sand, the nature and quality of the binder also influence this property (CINCOTTO et al, 1995). In addition, Pereira et al (1999) believe that consistency is also directly linked to bond strength.

In terms of consistency, mortars can be classified qualitatively into (CINCOTTO et al, 1995):

• Dry mortars: these are mortars that require the application of energy to form them into their final shape, where the paste fills the voids between the grains;

• Plastic mortars: mortars that reach their final form with a little effort, where the paste forms a thin film and acts as a lubricant on the surface of the aggregate grains;

• Fluid mortars: these are mortars that flow and level themselves, requiring no effort other than the force of gravity to apply them, in which case the grains are immersed in the paste.

According to Selmo (1989) apud Silva (2006), in mortars with a plastic or fluid consistency, water exudation can occur. This property also interferes with workability, requiring frequent mixing to homogenize the material and can interfere with the mortar's ability to adhere when applied to the substrate.

The flow table prescribed by NBR 7215 (ABNT, 1996) and the test procedure for determining the consistency index prescribed by NBR 13276 (ABNT, 2005) are used to assess consistency. However, despite its widespread use, this test has been widely criticized by several authors who claim that the table is not sensitive enough to measure mortar rheology (GOMES et al, 1995; YOSHIDA and BARROS, 1995; CAVANI et al, 1997; JOHN, 2003; BAUER et al, 2005).

According to Bauer et al (2005), despite the existence of sophisticated equipment that allows a broader assessment of the rheological behavior of mortars, including the fundamental parameters (viscosity and flow tension), it is not common in mortar technology laboratories, due firstly to its high cost and secondly to some operational difficulties.

Retention of consistency can be defined as the property of mortar to maintain its consistency after coming into contact with a substrate. This property is of fundamental importance for mortars used to lay masonry and ceramic tiles, and basically depends on the mortar's good water retention.

3.1.5.1.3 Cohesion and thixotropy

The cohesion of a mortar can be defined as the property of the mortar to keep its constituents homogeneous without segregation, through the physical forces of attraction between the solid particles of the mortar and the chemical bonds of the binder. Mortars need special additions and suitable binders to achieve good cohesion. Despite the existence of various additives, the use of hydrated lime is still the most widely used way of achieving this property (CINCOTTO et al, 1995).

In addition, according to the same authors, lime provides the internal cohesion of the mortar by reducing the surface tension of the binder and the adhesion to the aggregate, as well as influencing the workability, consistency and plastic stability of the mortars.

Thixotropy is also a property related to cohesion. It can be defined as a change in viscosity when the mass is stirred, causing it to move from a solid or paste state to a gel (CALHAU, 2000). In the case of mortar, the gel state refers to the cohesive mass of binder in the paste which becomes denser after hydration (CINCOTTO et al, 1995).

Thixotropy is required, for example, in mortar for laying ceramic tiles and restoration mortars. Thixotropic mortars require low energy to change their shape, but once altered they can maintain it even under the action of gravity. To achieve this, additives based on polymers and mineral additions such as fly ash, microsilica and rice husk ash can be used.

3.1.5.1.4 Plasticity

Plasticity is the property that allows mortar to deform and retain certain deformations after the stresses to which it has been subjected have been reduced. According to Cincotto et al (1995), plasticity together with consistency is the property that characterizes workability and is basically influenced by the content of trapped air, the nature and content of binders and the mixing intensity of mortars. The plasticity of a mortar is directly linked to its consistency, cohesion and water retention.

The appropriate plasticity for each mixture, according to the purpose and form of application of the mortar, requires an optimum amount of water which means an optimum consistency, this being a function of the proportioning and nature of the materials (CASCUDO et al, 2005).

3.1.5.1.5 Water retention

"Water retention is the ability of fresh mortar to maintain its consistency when subjected to stresses that cause loss of mixing water, either through evaporation, suction or absorption by a surface" (TRISTÂO, 1995, page 10), or simply the ability of mortars to retain mixing water against suction from the substrate or evaporation (BAiA and SABATTINI, 2000).

Water retention determines the mortar's handling conditions and influences its properties in the hardened state, such as drying shrinkage, adhesion and final mechanical strength, as it determines the conditions for cement hydration and lime carbonation, which are responsible for the hardening of mixed lime and cement mortars.

The water retention capacity of mortars varies depending on the absorption potential of the substrate, and can also be influenced by:

- Specific area of the constituent materials;

- Pre-maturation of lime mortars (period in which the lime paste or mortar is left to rest before application);

- Nature of the lime, where dolomitic lime has better characteristics than calcitic lime (CINCOTTO et al, 1995);

- Lime/cement ratio in the mix;

- Aggregate/binder ratio of the mix.

Due to their specific area, binders are mainly responsible for the water retention capacity of mortars. According to Roman et al (1999), poor water retentivity can result from poor aggregate granulometry, very large aggregates, insufficient mixing or choice of mix can increase retentivity. The use of pozzolanic material or the addition of more water and a longer mixing time can further increase this retentivity.

Mortars with high water retention allow the cement to hydrate without losing its plasticity, while mortars with low water retention make handling difficult and can cause the substrate to expand (MÜLLER et al, 1999).

As with workability, the factors that influence water retention are the characteristics and proportions of the mortar's constituent materials. The presence of lime and additives can improve this property (BAiA and SABATTINI, 2000).

3.1.5.1.6 Specific mass and incorporated air content

According to Roman et al (1999), the specific mass corresponds to the ratio between the mass of the material and its volume and can be absolute or relative. When determining the absolute specific mass, the voids in the volume of the material are not taken into account. In the case of relative mass, also known as unit mass, the voids present are taken into account.

According to the same author, both the unit mass and the air content will affect other properties of the mortar in its fresh state, such as workability. A mortar with a lower specific mass and a higher air content has better workability.

Air content is defined as the amount of air present in a certain volume of mortar. As the air content increases, the relative specific mass decreases (ROMAN et al, 1999).

The voids present in the mortar are actually air incorporated or trapped during the mixing and application process, or spaces left after the evaporation of excess water (capillary voids). Air content has an influence on the workability and adhesion strength of coatings, so the use of air-entraining additives should be limited. Improper use of these additives can cause damage to the mechanical strength and adhesion of the mortar (ROMAN et al, 1999).

3.1.5.1.7 Exudation

Exudation is the phenomenon of part of the mixing water separating from a fresh mortar kept at rest without any kind of vibration or shock.

It is known that water is the lightest component of the mortar mix, and consequently exudation is a form of segregation, because suspended solids tend to settle under the action of gravity. Exudation results from the inability of the component materials to retain all the water in the mixture in a dispersed state, while the heavier solids are settling (MEHTA and MONTEIRO, 1994).

Exudation generally occurs in mortars with a plastic to fluid consistency. This property interferes with the workability of mortars, as it requires constant re-mixing to homogenize them and according to Selmo (1989) apud Tristao (1995) it can interfere with adhesion when cast against the substrate.

3.1.5.1.8 Initial adhesion

It can be defined as the property related to the mechanical phenomenon that occurs on porous surfaces, by the anchoring of the mortar to the base, through the entry of the paste into the pores, recesses and protrusions, followed by the progressive hardening of the paste, as illustrated in Figure 5 (BAiA and SABATTINI, 2000).

Source: BAiA and SABATTINI, 2000.

Figura 5 - Anchoring the mortar to the base.

Initial adhesion is the property that will characterize the future behavior of the substrate/ mortar combination in terms of the performance resulting from adhesion (CINCOTTO et al, 1995), i.e. this property is directly linked to the adhesion that the mortar will have to the substrate in the hardened state.

Initial adhesion basically depends on the properties of the mortar in its fresh state; the characteristics of the substrate to be applied, such as porosity, roughness, cleaning conditions and the effective contact surface between the mortar and the substrate. In order to obtain adequate initial adhesion, as shown in Figure 6, the mortar must be workable and retain adequate water to absorb the substrate on which it will be applied (BAÎA and SABATTINI, 2000).

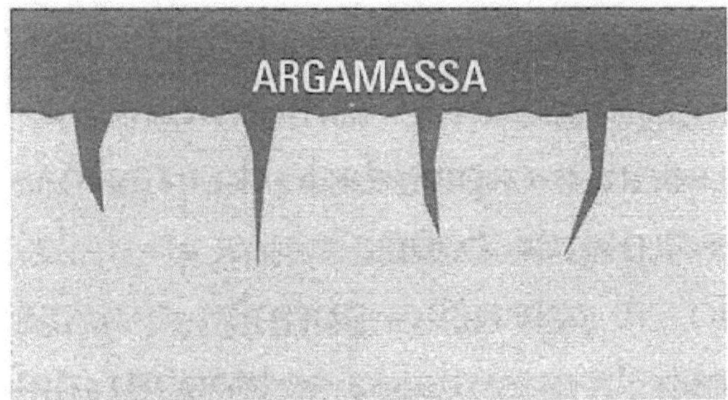

Source: BAiA and SABATTINI, 2000.

Figura 6 - Adequate initial adhesion.

The roughness and cleanliness of the substrate, which must be free of dust, loose particles and oil, are considered essential factors for good initial adhesion of the mortar (SILVA, 2006).

3.1.5.1.9 Hardening time

The hardening time of mortars depends on the hydration of the cement, i.e. the chemical reaction between cement and water. If hardening takes place too quickly, it will cause problems when laying or finishing the masonry cladding; if it is too slow, it will delay construction. Very high temperatures tend to accelerate the hardening of mortars and low temperatures cause it to slow down (ROMAN et al, 1999).

According to the same author, the more homogeneous the mixture, the better the cement will spread, and its contact with water will consequently accelerate the hardening process.

3.1.5.1.10 Plastic shrinkage

According to Baia and Sabattini (2000), plastic shrinkage occurs due to the evaporation of the mortar's mixing water and also due to the hydration and carbonation reactions of the binders. Shrinkage can end up causing cracks to form in the coating.

Plastic shrinkage occurs in the early stages of mortar production and is linked to the internal reactions of the cement paste, occurring before the cement sets (AÏTCIN, 2000). This shrinkage is irreversible and does not occur at advanced ages. The loss of water from the cement paste is the main cause of shrinkage (BARBOSA, 2005).

Cracks can be harmful or not (microcracks). Damaging cracks allow water to seep through the coating when it has hardened, compromising its watertightness (BAiA and SABATTINI, 2000).

The same authors also point out that the factors that influence this property are: the characteristics and proportioning of the mortar's constituent materials; the thickness and application interval of the layers; respect for the setting and leveling time. Unlike mortars with a high cement content (strong mortar), mortars with a low cement content (weak mortar) are less subject to the stresses that cause damaging cracks to appear during drying, as well as cracks and possible detachment of the mortar when it has hardened (Figure 7).

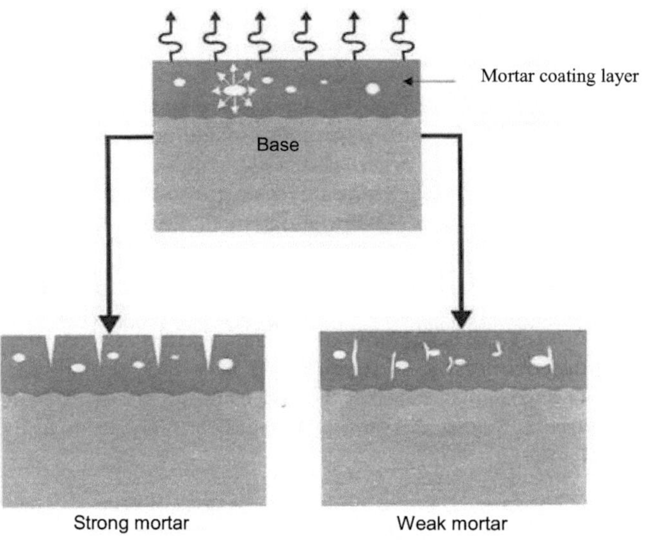

Source: BAiA and SABATTINI, 2000.

Figure 7 - Strong mortar and weak mortar.

3.1.5.2 Properties in the hardened state

3.1.5.2.1 Mechanical resistance

Mechanical resistance can be defined as the property of mortars to withstand mechanical actions of different kinds, due to surface abrasion, impact and hygroscopic contraction (BAIA and SABATTINI, 2000).

After hardening, regardless of the type of application, mortars will always be subjected to some kind of mechanical stress. Laying mortars are subjected to compression and coating mortars to surface abrasion, impact, shear stresses resulting from substrate movement and thermal/hygrometric variations.

Mechanical strength basically depends on the consumption and nature of the aggregates and binders in the mortar used and on the execution technique, which seeks to compact the mortar during its application and finishing. Mechanical strength increases as the proportion of aggregate in the mortar decreases and varies inversely with the water/cement ratio of the mortar (BAIA and SABATTINI, 2000). However, other factors such as temperature and humidity can have a significant impact on the mechanical strength of mortar (KOLIAS, 1994).

The mechanical properties of the mortar are very important for the compressive strength of the masonry, since the mechanism of rupture of the wall is directly linked to the joint/unit interaction (ROMAN, 1991).

The mortar must be strong enough to withstand the stresses to which the wall will be subjected. It is important to note that a high compressive strength of the mortar does not necessarily mean a stronger wall, i.e. a better structural solution (ROMAN et al, 1999).

3.1.5.2.2 Elasticity

Elasticity or resilience is the property of mortar to withstand stresses without breaking, without damaging cracks and without losing adhesion (BAiA and SABBATINI, 2000), returning to its initial dimensions when the stresses imposed on it cease.

However, this meaning is extended, in the case of mortars, to such a state of plastic deformation in which rupture occurs in the form of microscopic cracks or non-damaging capillaries (SABATTINI, 1986).

Cracks are the result of stress relief caused by the deformation of the base. They are considered harmful when they allow water to percolate into the coating. The appearance of damaging cracks compromises adhesion to the substrate, watertightness of the surface and durability of the building (BAÎA and SABBATINI, 2000).

The mortar's ability to absorb deformations can be assessed using the modulus of elasticity, which can be obtained using the static or dynamic method. The lower the value of the modulus, the greater the material's capacity to absorb deformations (SILVA, 2006).

Mortars must have this property in most of their applications, and it is of fundamental importance for mortars for laying and coating masonry (OLIVEIRA, 2001).

3.1.5.2.3 Retraction

Shrinkage is a process suffered by mortars, especially in the early stages. In the hardened state, shrinkage occurs shortly after the mortar hardens, as a result of the chemical reaction of the binders (hydrated lime and Portland cement) and the removal of water adsorbed in the hydration products during the drying process (KOPSCHITZ et al, 1997).

According to Sanchéz et al (1997), shrinkage can be defined simply as the process of volume reduction suffered by cementitious matrices during and after hardening, when exposed to air. The authors also point out that this phenomenon generates internal stresses in mortars which can cause anything from a reduction in volume to cracking of the material. The propagation of cracks in a heterogeneous material such as mortar and concrete is influenced by its components (DETRICHE and RAMODA, 1989).

Studies carried out by Fiorito (2003) show that 35% to 45% of total shrinkage has already occurred at seven days, 50% to 60% at twenty-eight days and 80% to 95% at one hundred and twenty days.

According to Barbosa (2005), the phenomenon of shrinkage can be classified as follows:

•   The physical state of the mixture (plastic shrinkage and shrinkage in the hardened state);

•   Nature (water loss shrinkage, hydration shrinkage or chemical shrinkage, autogenous or endogenous shrinkage, carbonation shrinkage, thermal shrinkage);

•   The degree of restriction (free retraction and restricted or impeded retraction);

•   Permanence (retraction to the permanence of the phenomenon and irreversible or permanent retraction).

Against this backdrop, we will now look only at the shrinkage that can occur during the hardened state of mortars.

The removal of water from concrete or mortar, both kept in unsaturated air, causes shrinkage due to water loss (NEVILLE, 1997).

Regarding hydration shrinkage or chemical shrinkage, Bastos (2001) apud Barbosa (2005) reports that when cement is hydrated, the total volume of hydrates is less than the initial absolute volume (absolute volumes of water + anhydrous cement). This contraction phenomenon was discovered by Lê Chatelier and is named after him. After complete hydration, the concentration of Lê Chatelier is around 10% of the initial volume of the paste. This phenomenon occurs from just before setting and continues throughout hydration, being much more intense in the first few days.

Autogenous or endogenous shrinkage is a combination of the chemical shrinkage of hardened mortar and a phenomenon resulting from it: the reduction of free water in the capillary pores. This occurs in a closed system, with the capillary water being removed from the paste by the hydration of the not yet hydrated cement. It occurs without the exchange of moisture with the external environment (BARBOSA, 2005).

According to Neville (1997), carbonation shrinkage probably occurs due to the dissolution of

Ca(OH) crystals$_2$, under tension due to hydraulic shrinkage, and the deposition of $CaCO_3$ in spaces not subject to tension. If carbonation occurs after the C-S-H dehydration stage, carbonation shrinkage also results. Briefly, Barbosa (2005) defines carbonation shrinkage as the consequence of the reaction of CO_2 from the air with hydrated cement compounds in the hardened paste with an increase in mass.

The first hours of cement hydration are accompanied by a significant release of heat. The decrease in temperature causes shrinkage, which can occur in the cement-based component in two different ways: after the peak temperature evolution due to the accumulation of hydration heat and due to the thermal variation of the environment (BARBOSA, 2005).

According to the same author, in terms of the degree of restriction, mortars can undergo free shrinkage or restricted (impeded) shrinkage. Free shrinkage occurs in a test specimen or isolated structural element, i.e. it is free to shrink; in the case of restricted (impeded) shrinkage, it can be said that various forms of restriction can arise against the free occurrence of shrinkage, such as restriction due to the presence of aggregate in mortars and concretes, restriction due to the adherence of the paste, mortar or concrete to a base, restriction due to the adherence of concrete to reinforcement, in the case of reinforced concrete. However, with regard to the permanence of shrinkage, mortars in the hardened state can undergo reversible shrinkage, i.e. part of the shrinkage can be recovered by wetting the cement paste after drying.

Shrinkage has a significant influence on the watertightness and durability of coatings. The factors that affect the shrinkage of mortars are the binder content, the volume of water, the grain size of the aggregates and the environmental conditions (OLIVEIRA, 2001).

Another factor relevant to shrinkage is the curing conditions. If the water evaporates before the binder has set, shrinkage can be very high. When concrete or mortar is kept at a constant relative humidity, shrinkage is much faster in the early stages, reaching around 50% of total shrinkage after just seven days of conditioning (THOMAZ, 2003).

According to Bortoluzzo et al (1999), mixes richer in cement have higher shrinkage values.

The water/cement ratio is the most influential factor in shrinkage, more so than the influence of cement consumption itself (THOMAZ, 2003). In addition to all the internal factors mentioned above, according to BRS (1970) apud Thomaz (2003), the geometric shape of the piece has a significant influence on the amount of shrinkage. The greater the ratio of the exposed area of the piece to the volume of the piece, the greater the shrinkage to be developed.

In coating mortars, in addition to intrinsic factors, factors such as adherence to the base, the number of layers applied, the thickness of the layers, the time elapsed between the application of one layer and another, the rapid loss of water during the hardening process due to intensive ventilation and/or sunlight, among others, can cause shrinkage (THOMAZ, 2003).

3.1.5.2.4 Adherence

Adhesion is the ability of the mortar to remain fixed to the substrate by resisting the normal and tangential stresses that arise at the substrate- mortar interface. It is the result of the mortar's tensile strength, shear strength and adhesion extension (BAÌA and SABBATINI, 2000).

According to Taha and Shrive (2001) apud Carvalho Jr. et al (2005), it is widely accepted that adhesion to masonry develops according to two mechanisms:

•	Chemical adhesion: adhesion strength comes from covalent forces or Van der Waals forces developed between the masonry unit and the cement hydration products;

•	Mechanical adhesion: formed by the mechanical interlocking of cement hydration products transferred to the surface of the pores in the masonry unit.

Authors such as Cincotto et al (1995) and Roman et al (1999) report that adhesion is significantly influenced by the characteristics and conditions of the substrate - roughness, porosity, water absorption, mechanical resistance and surface texture; type of binder used; aggregate granulometry; initial adhesion; water retention; consistency; air content of the mortar; labor; temperature and relative humidity. In addition, Baia and Sabbatini (2000) point out that the surface cleanliness of the substrate also contributes to the good performance of this property.

Mixed mortars (cement and lime), in general, due to their plasticity and water retention, allow irregularities in the base to be filled in and have a favorable effect on adhesion.

Authors such as Carasek (1997); Rocha and Oliveira (1999) and Pereira et al (1999) conclude that the cement content influences the adhesion strength of mortars, i.e. as cement consumption increases, higher adhesion strength values are obtained.

In their studies, Carasek (1997) and Pereira et al (1999) concluded that wetting the ceramic substrate does not contribute to gains in mortar bond strength. However, Pereira et al (1999) believes that the curing of the coating has a major influence.

3.1.5.2.5 Permeability

The behavior of building materials in relation to humidity is largely dependent on their water absorption (KRUS et al, 1997).

Permeability is related to the passage of water through the mortar layer, which is a porous material and allows water to percolate both in liquid and vapor form (BAÌA and SABBATINI, 2000). Where the mortar has already hardened, water can penetrate it by means of pressure infiltration, capillarity or diffusion of water vapor.

The permeability of a mortar basically depends on the amount and type of binder used, the particle size of the aggregate and the characteristics of the substrate. Portland cement, when used in adequate proportions, can greatly reduce the permeability of a mortared coating, but if used in very high levels it can cause cracking due to hydraulic shrinkage, thus compromising the permeability of the mortar.

The cladding must be watertight, preventing it from percolating. However, it must be vapor-permeable in order to encourage the drying out of moisture from infiltration (such as rainwater, for example) or from the direct action of water vapor, especially in bathrooms (BAiA and SABBATINI, 2000).

Permeability is not a simple function of porosity; it also depends on the size, distribution and continuity of the pores (NEVILLE, 1997). According to Kolias (1994), porosity is an important factor when it comes to the durability of mortars.

3.1.6 Durability

No material is inherently durable; as a result of environmental interactions, the microstructure and, consequently, the properties of materials change over time. A material is considered to have reached the end of its useful life when its properties, under given conditions of use, deteriorate to such an extent that the continued use of this material is considered unsafe or uneconomical (MEHTA and MONTEIRO, 1994).

Therefore, the durability of a coating can be defined as the property of the period of use of this coating, resulting from its properties in the hardened state, and which reflects the performance of the mortars in the face of the actions of the external environment over time (BAiA and SABBATINI, 2000).

Inadequate durability is manifested by deterioration that can be caused by external factors or by internal causes within the material itself. The different forms of action can be physical, chemical or mechanical (NEVILLE, 1997).

According to Sabbatini (1986), mortars can have their integrity compromised by a series of factors, including drying shrinkage, absorption of rainwater, freezing temperatures, thermal shock, atmospheric corrosive agents and aggressive biological agents. In addition to these factors, according to Baia and Sabbatini (2000), excessive thickness and lack of maintenance can also compromise the durability of the material.

3.1.7 Microstructural analysis

Effective methods for determining the phases of materials are based on microscopy, X-ray diffraction and chemical analysis (TAYLOR, 1990).

All solid materials contain large numbers of imperfections or deviations from crystalline perfection. Microscopic examination is an extremely useful tool in the study and characterization of these materials. Microscopy can be related to the various characteristics and properties of materials (CALLISTER, 2002).

Microstructure is the fine structure of a material, observable with the aid of a microscope. In the study of construction materials, it is known that there is a relationship between the manufacturing process, the microstructure and the properties. In view of this, image analysis is helping to improve the level of knowledge of these materials and is used as an appropriate tool to assess the morphological changes that occur during the process and its use (GONDIM, 2005).

3.1.8 Classification of mortars

According to NBR 13530 (ABNT, 1995), mortars are classified as follows (Table 2):

Type	Ranking criteria
Aerial mortar	
Hydraulic mortar	Nature of the binder
Lime mortar	
Cement mortar	Type of binder
Cement and lime mortar	
Simple mortar	
Mixed mortar	Number of binders
Additive mortar	
Improved adhesion mortar	
Adhesive mortar	
Permeability-reducing mortar	Special properties
Radiation protection mortar	
Water-repellent mortar	
Thermo-insulating mortar	
Plastering mortar Plastering mortar Plastering mortar	Coating function
Centrally dosed mortar Site-prepared mortar Industrialized mortar Semi-ready mortar mix	Method of preparation or delivery

Source: ABNT (1995).

Table 2 - Classification of mortars according to NBR 13530.

The mortars usually used in construction are lime mortars, cement and sand mortars and cement, lime and sand mortars, better known as mixed mortars (SILVA, 2006).

3.1.8.1 Lime mortar

Lime mortar is made up of lime, fine aggregate and water. The lime paste fills the voids between the grains of the fine aggregate, improving plasticity and water retention. Lime mortar is usually called intermediate mortar, because when virgin lime is used, this type of mortar is used to mature the lime so that the cement can be mixed in later (SILVA, 2006).

3.1.8.2 Cement mortar

This type of mortar is basically composed of cement, fine aggregate and water. It acquires high mechanical resistance in a short time, but has poor workability and low water retention, as well as being more prone to shrinkage. This type of mortar is rarely used as a coating; its main use is to make plaster to be applied to masonry walls and concrete structures to increase the adhesion strength of the mixed mortar coating (SILVA, 2006).

3.1.8.3 Mixed mortar

Mixed mortar is basically composed of cement, lime, fine aggregate and water. According to Sabbatini (1986), the English used a 1:3 ratio (cement+lime:dry sand) by volume as their basic mix, because they assumed that with this ratio the voids in the sand would be filled by the binder. Thus, by changing the relative proportions of lime and cement, workability is more or less guaranteed, regardless of the changes. This ratio is widely used in Brazil, such as the traditional 1:2:8 (cement:lime:fine aggregate) volume mixes for internal cladding, 1:1:6 for external cladding and 1:0.5:4.5 for laying structural masonry units.

3.2 QUARRY WASTE

3.2.1 Definition

This is the material resulting from the crushing of any type of rock, i.e. the production of coarse aggregate, generally with a diameter of less than 2.4 mm, sold dry. It usually has low added value and is therefore considered waste.

3.2.2 Sourcing and characteristics

The use of the waste generated by the crushing process is not new. According to Nuget (1979) apud Menossi (2004), this material has been used for centuries in various countries, such as the United States and Canada.

The natural stone sector (extraction and processing of stone) does not give rise to hazardous waste, but the type of processing results in the production of fairly large quantities of waste, particularly solids during extraction, which will cause considerable impacts and constitute a nuisance and burden for companies.

According to Fujimura et al (1995) apud D'Agostini and Soares (2003), quarries generate significant amounts of waste during the rock processing process, especially during the crushing process. It is estimated that between 10% and 15% of the volume of gravel produced is the total waste generated in the crushing process, which results in several million tons accumulated annually. Figure 8 illustrates the storage of quarry waste.

Source: MENOSSI, 2004.

Figure 8 - Example of how quarry waste is stored in the company yard.

Table 3 shows the most common classification adopted by Brazilian mining companies for rock crushing products (PISSATO and SOARES, 2001).

Numbered gravel		Size	
		Minimum (mm)	Maximum (mm)
Waste	Stone dust		< 0,075
	Gravel	0,075	4,8
Brita 1		4,8	12,5
Brita 2		12,5	25
Brita 3		25	50
Brita 4		50	76
Brita 5		76	100
Rachao		> 100	

Source: FAÇO Manual, modified by MENDES (1999) apud PISSATO and SOARES (2001).

Table 3 - Particle size classification of rock crushing products.

If considered suitable for the preparation of mortars, quarry waste would be transformed from waste into a mining by-product, thus promoting a use for the large volume of fines that can generate various environmental impacts if not properly managed (D'AGOSTINO and SOARES, 2001).

Usually, quarry waste has a cubic shape, prominent rough and angular surfaces and a mineralogical composition composed basically of quartz, potassium feldspar (microcline), plagioclase and micas

(PISSATO and SOARES, 2001).

3.2.3 Use in cementitious matrix

Research carried out by D'Agostino and Soares (2001) shows that quarry waste is suitable for preparing mortars. According to the authors, the material has a good grain size distribution, which allows for better filling of voids left by larger grains and, therefore, mortars can be produced with smaller amounts of cement.

According to Neves et al (2002), although concrete dosed with quarry waste requires a little more water, which is insignificant in a large-volume mix, since for a mix of 1 m^3 , it requires only 2 to 4 liters more than its standard mix (without the addition of quarry waste), concrete dosed with quarry waste consumes less cement and has greater mechanical strength than standard concrete.

The research carried out by Silva et al (2007) shows that:

•   Mortars made with quarry waste have lower levels of incorporated air than mortars made with natural sand, possibly due to the higher content of powdery material present in quarry waste;

•   With regard to bulk density in the hardened state, mortars made with quarry waste have a higher value than mortars made with natural sand. This property can be used to characterize mortars in terms of the presence of voids and mechanical strength.

According to D'Agostino and Soares (2003), mortar produced with quarry waste, after curing for 90 days, has a higher compressive strength, which is more than double the result obtained with mortar produced with alluvial sand. Paes et al (1999) state that the higher the quarry waste content, the greater the incidence of cracking, despite the reduction in unit water. This cracking is probably due to shrinkage, which can be explained by the grain size of the sand and the quarry residue. The sand acts as a "structure" to prevent the mortar from shrinking, and when it is replaced at different levels by finer materials, it facilitates the occurrence of cracks, since the smaller the diameter of the aggregates, the greater the shrinkage (PAES et al, 1999).

According to Silva et al (2005), replacing natural sand with quarry waste is feasible in the manufacture of mortar for laying and coating.

Research carried out by Angelim et al (2003) concludes that the addition of quarry waste to mixed mortars improves the mechanical strength, surface hardness and resistance to abrasion. This is due to the low water/cement ratio, the high compactness of the hardened mortars and the possible integration of the quarry residue into the hydrated cement paste.

When natural sands are replaced by quarry waste in mortars, they become heavier, thus affecting the density of the mass in the fresh state, making them unsuitable for use as ceiling and wall coverings (TRISTÂO et al, 2005).

3.3 PORCELAIN TILE POLISHING WASTE

3.3.1 Definition

Porcelain tile polishing residue can be defined as a mixture of ceramic material from polishing porcelain tiles and abrasive material released during the polishing process.

3.3.2 Sourcing and characteristics

polished porcelain tiles are the most commercially produced porcelain tiles, requiring a polishing stage during production in order to level, remove scratches and defects and give a shine to the surface of the final product. As a result, the porcelain tile industry generates a large amount of waste during the polishing process. This creates a new cost for producers in the sector, who need to manage and dispose of this waste properly (MARQUES et al, 2007).

The porcelain tile polishing process takes place in equipment that has several polishing heads basically made up of abrasive materials, which in direct contact with the tiles at high speed and controlled rotation in the presence of water, polish them, gradually using a finer grain size until the desired result is achieved (SILVA, 2005).

According to the definition already described in section 3.3.1, porcelain tile polishing residue is made up of a mixture of ceramic material from polishing porcelain tiles and abrasive material released during the process, resulting in a white powder. The abrasive material (Figure 9) used during polishing is generally made up of diamond or silicon carbide particles agglomerated with magnesium chloride-based cements (MARQUES et al, 2007).

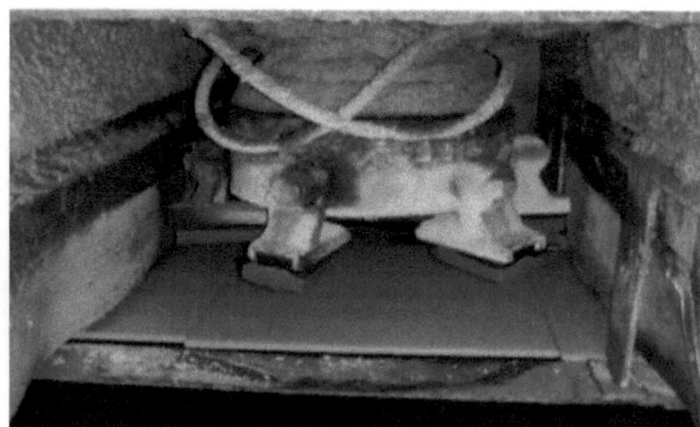

Source: SOUZA, 2007.

Figure 9 - Polishing heads with abrasive bricks.

The chemical composition of porcelain tile polishing waste is basically made up of the porcelain tile itself and the waste generated by the polishing tools. This discarded material is usually stockpiled by the industry (Figure 10) for later use in backfilling quarries exploited to extract the raw material used to manufacture the porcelain tiles themselves (SOUZA, 2007).

Source: SOUZA, 2007.

Figure 10 - Storage of porcelain tile polishing residue.

In this sense, reuse and recycling are the most suitable conditions for handling the vast majority of industrial waste, such as porcelain polishing waste, reducing costs, preserving non-renewable natural resources, reducing the load of pollutants released into the environment (BERNARDIN et al, 2006) and reducing the environmental impacts and harmful effects on human health caused by the indiscriminate disposal of waste in nature.

As polished porcelain tiles are the most widely sold on the market, there is a permanent production of waste from their production. According to information from Mr. Mauro Palma, manager of the Elizabeth Porcelanatos industry, located in the state of Paraiba, the production of porcelain tile polishing waste is between 30 and 35 tons per day (SOUZA, 2007). In the Cecrisa industry, located in the state of Minas Gerais, approximately 20 tons of waste are produced daily (SILVA, 2005).

Porcelain tile polishing waste is fine-grained and has a varied chemical composition (SOUZA, 2007), with a high concentration of silica (SiO_2), alumina ($Al\ O_{22}$) and magnesium oxide (MgO) in its composition (SILVA, 2005).

Also, according to Silva (2005), the alkali content found in the chemical analysis is lower than that allowed in cements. According to NBR 12653 (ABNT, 1992), which classifies pozzolans into classes according to their origin and minimum requirements in terms of their composition, the porcelain tile polishing residue appears in class N as a calcined clay with a SiO2 + Al2O3 + FeO3 content of over 70%. The SO oxide$_3$ is below 4%, the alkali $Na_2\ O$ is just above 1.5% and the loss on ignition is below the maximum percentage of 10%. Therefore, according to the chemical analysis carried out, the porcelain tile polishing residue can be classified as a pozzolan, although it is necessary for the silica to be in an amorphous state.

43

X-ray diffraction is very useful for obtaining the spectrum with important information for characterizing the waste. The intensity of the characteristic peaks of each mineral is proportional to its quantity. Phase identification is done by comparing the spacings (or diffraction angles) of the sample with known data for crystalline materials available in a database (SILVA, 2005).

Structural analysis makes it possible to identify the phases present, the structure and the microstructure of a material. Figures 11, 12, 13 and 14 below clearly illustrate the structural characterization of porcelain tile polishing residue.

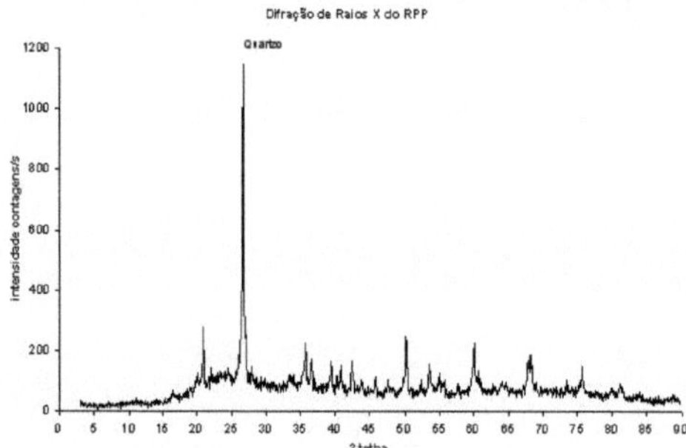

Source: SILVA, 2005.

Figure 11 - X-ray diffraction of porcelain tile polishing residue.

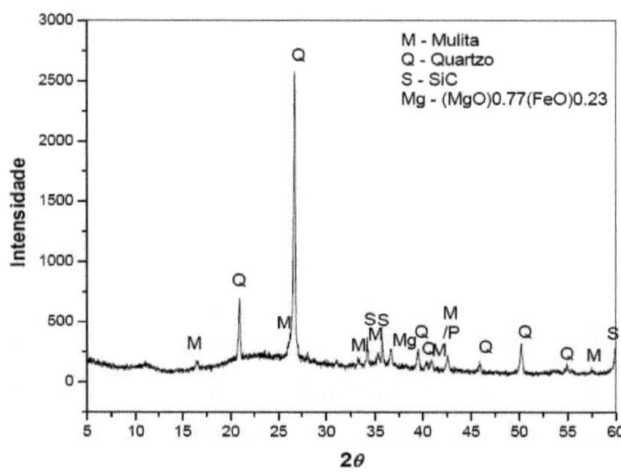

Source: MARQUES et al, 2007.

Figure 12 - X-ray diffraction of RPP.

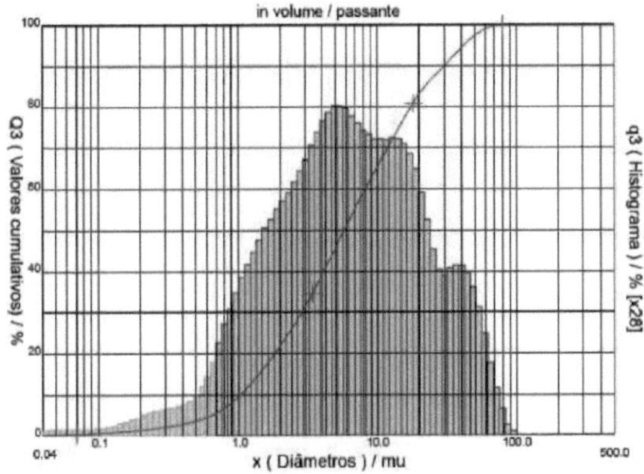

Source: MARQUES et al, 2007.

Figure 13 - RPP particle size distribution.

Source: SILVA, 2005.

Figure 14 - Residue particle from polishing retro mirror porcelain tiles.

According to Silva (2005), the specific mass of porcelain tile polishing residue is approximately 2.49 g/cm^3 .

3.3.3 Use in cementitious matrix

In the literature, there are not many publications that deal directly with this subject. Most of the research carried out on the use of porcelain tile polishing residue involves its use in the manufacture of new ceramic products. Authors such as Silva (2005) and Souza (2007) used porcelain tile polishing residue as an addition to concrete in their research.

According to Souza (2007), the use of porcelain tile polishing residue in concrete provides great environmental benefits and also improves mechanical strength and durability.

Porcelain tile polishing residue, when mixed with CP V cement, shows pozzolanic activity (SILVA, 2005).

According to the same author, concrete mixtures with the addition of porcelain tile polishing residue have a better consistency and less exudation. In terms of axial compressive strength, concrete with the addition of porcelain tile polishing residue performed better than concrete without the addition of the residue.

With regard to durability aspects, it was found that the presence of porcelain tile polishing residue results in lower porosity and absorption in the concrete. The presence of porcelain tile polishing residue in concrete does not alter its thermal conductivity (SOUZA, 2007).

CHAPTER 4

EXPERIMENTAL PROGRAM

4.1   DEFINING MORTAR MIXES

The mortars used in this experimental study were mixed laying mortars with a 1:1:6 (cement:lime:sand) mix, by volume, commonly used in the execution of sealing and structural masonry, this mix can also be used as a mortared coating on the façades of buildings over 20 floors; mixed laying mortars with a specific structural function, with a 1:0.5:4.5 (cement:lime:sand), by volume, commonly used in structural masonry buildings over 4 (four) floors. The aim was to use mortars with traits commonly used on construction sites, thus portraying part of the reality of Brazilian buildings.

4.2   MORTAR FORMULATION

All the mortars used in this experimental study were made in the Civil Construction Materials Laboratory at CEFET/RN.

These mortars were formulated by replacing different levels of cement with limestone quarry waste (RPPC) and porcelain tile polishing waste (RPP). As shown in Table 4, a total of 16 (sixteen) mortars were formulated with substitutes, considered mortars with residues, and 2 (two) mortars without additions, considered reference mortars (REF), which served as a comparative parameter.

Mortar mix	Mortar	Residue content (%)	Waste	Nomenclature
1:1:6	A1	0	REF*	A1/0/REF
1:1:6	A1	5	RPPC	A1/5/RPPC
1:1:6	A1	10	RPPC	A1/10/RPPC
1:1:6	A1	20	RPPC	A1/20/RPPC
1:1:6	A1	30	RPPC	A1/30/RPPC
1:1:6	A1	5	RPP	A1/5/RPP
1:1:6	A1	10	RPP	A1/10/RPP
1:1:6	A1	20	RPP	A1/20/RPP
1:1:6	A1	30	RPP	A1/30/RPP
1:0,5:4,5	A2	0	REF*	A2/0/REF
1:0,5:4,5	A2	5	RPPC	A2/5/RPPC
1:0,5:4,5	A2	10	RPPC	A2/10/RPPC
1:0,5:4,5	A2	20	RPPC	A2/20/RPPC
1:0,5:4,5	A2	30	RPPC	A2/30/RPPC
1:0,5:4,5	A2	5	RPP	A2/5/RPP

1:0,5:4,5	A2	10	RPP	A2/10/RPP
1:0,5:4,5	A2	20	RPP	A2/20/RPP
1:0,5:4,5	A2	30	RPP	A2/30/RPP

* reference mortar (without residue). Table 4 - Mortar formulation.

4.3 EXPERIMENTAL PLANNING

After defining the mortar traits and formulations, the materials used in the mortar dosages were characterized, followed by the conversion of the traits from volume to mass. After that, the mortar preparation was described, and then the mortar characterization was carried out. Figure 15 shows the schematic diagram of the steps taken to develop the experimental program.

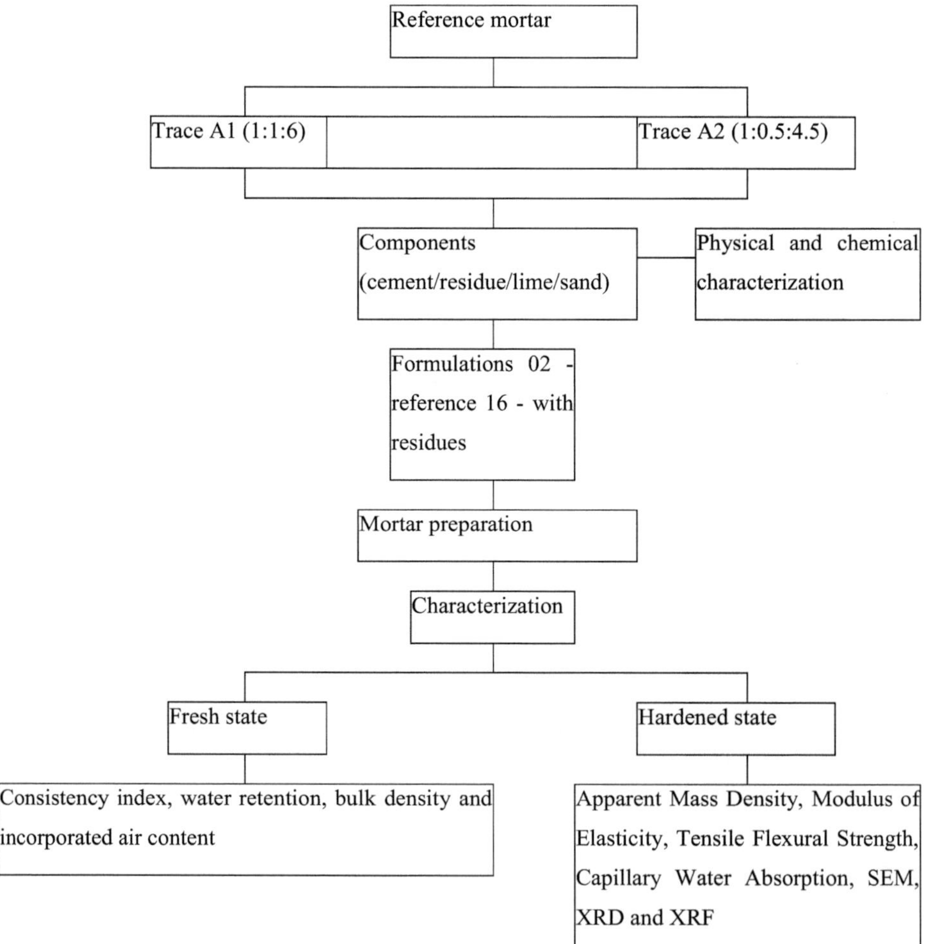

Figure 15 - Schematic diagram of the experimental procedure.

4.4 CHARACTERIZATION OF STARTING MATERIALS

The mixed mortars were made using cement, lime, natural sand, limestone quarry waste and porcelain polishing waste.

4.4.1 Cement

The cement to be used in the research was Poty CP II F 32 Portland cement from the same batch supplied by a concrete batching company. The choice of

type of cement was because it is a cement that does not have pozzolan in its composition, and is widely used in works in the region.

The cement was collected from one of the concrete batching company's silos. It was collected from the container in which it was stored. This is a plastic container, internally lined with plastic bags and sealed so that it is completely protected from the weather.

The chemical, physical and mechanical characteristics were provided by the manufacturer (Annex A). The specific mass and unit mass were obtained from tests carried out at CEFET/RN's Civil Construction Materials Laboratory, in accordance with NBR 9776 (ABNT, 1987) and NBR 7251 (ABNT, 1982), respectively.

4.3.1 Cal

To carry out the research, we used CH I hydrated lime from the same batch, supplied by the Carbomil industry, a lime producer located in the state of Ceará. The lime was stored in a plastic container, internally covered with a plastic bag and sealed, thus protecting it from the weather.

The physical and chemical characterization of the lime was provided by the manufacturer (Appendix B). The specific mass and unit mass were obtained in accordance with NBR 9776 (ABNT, 1987) and NBR 7251 (ABNT, 1982), respectively. The tests were carried out at the Civil Construction Materials Laboratory at CEFET/RN.

4.3.2 Small aggregate

The fine aggregate, here called quarzosa sand of natural origin, was purchased from a quarry in the district of Igreja Nova, located in the municipality of Macaiba/RN. A visual inspection showed that it was free of impurities and had a uniform grain size.

Approximately 2 m^3 of medium sand was collected in burlap sacks to be used in making the mortar. All the material was dried in an oven and then placed in plastic containers with lids, covered internally with plastic bags to protect it from the weather.

The fine aggregate was characterized in terms of granulometry, according to NBR 7217 (ABNT,

1987) and NBR 7211 (ABNT, 1983); specific mass, according to NBR 9776 (ABNT, 1987) and unit mass, according to NBR 7251 (ABNT, 1982).

4.3.3 Waste

The residues used in this work were first sieved through a 2.4 mm mesh sieve in order to exclude any particles with larger diameters that were not part of their original composition. No preliminary processing was carried out on the waste, nor was the portion below 0.075 mm discarded, in order to verify the possibility of using these materials exactly as they are stored in the yards of their producers. Any further processing would entail additional costs to the process.

4.3.3.1 Limestone quarry waste

The rock quarry waste used in the research comes from a limestone crusher located in the municipality of Mossoró/RN.

After drying in an oven, the limestone quarry waste was stored in plastic containers, covered internally with a plastic bag and sealed, thus protecting it from the elements.

The limestone quarry waste was characterized in terms of specific mass, unit mass, laser granulometry, scanning electron microscopy, X-ray diffractometry and X-ray fluorescence.

The specific mass and unit mass were obtained through tests carried out at CEFET/RN's Civil Construction Materials Laboratory, in accordance with NBR 9776 (ABNT, 1987) and NBR 7251 (ABNT, 1982), respectively.

Laser granulometry was determined using a CILAS model 1064 device, located in the Petroleum Research Laboratory, part of the chemistry department at UFRN.

The micrographs were taken at the Institutional Scanning Electron Microscopy Laboratory - LIMEV - UFRN, using a Phillips model XL 30 - ESEM.

X-ray diffraction analysis was carried out using Shimadzu XRD-7000 equipment belonging to the UFRN Cement Laboratory. The X-ray diffraction experiments were carried out using Cu-Kα radiation, with an accelerating voltage of 30 kV and a current of 30 mA, scanning from 5° to 80°, at a scan speed of 2°/min.

X-ray fluorescence was carried out using the Shimadzu EDX 720 equipment, which is located in the Materials Characterization Laboratory at UFRN.

4.3.3.2 Residues from polishing porcelain tiles

The porcelain tile polishing residue was supplied by Indùstria Elizabeth Porcelanatos S/A, located in the state of Paraiba. This company specializes in the production of porcelain tiles.

All the material was dried in an oven and then stored in sealed plastic containers, lined internally with plastic bags, where it remained protected from the weather.

The porcelain tile polishing residue was characterized using specific mass, unit mass, laser granulometry, scanning electron micrography, X-ray diffractometry and X-ray fluorescence.

The specific mass and unit mass were obtained through tests carried out at CEFET/RN's Civil Construction Materials Laboratory, in accordance with NBR 9776 (ABNT, 1987) and NBR 7251 (ABNT, 1982), respectively.

The same equipment used to characterize the limestone quarry waste was used to carry out the laser granulometry, scanning electron microscopy, X-ray diffraction and X-ray fluorescence analysis of the porcelain polishing waste.

4.3.4 Water

The water used came from the local public water supply.

4.5 CONVERSION FROM TRACE VOLUME TO MASS

It can be said that the proportion of the constituent materials in mortars has a significant influence on their properties. In this way, the methodology used to measure the materials, initially measured in volume, and then converted into mass, after determining the unit masses of cement, waste, lime and ground aggregate, becomes relevant (CALHAU, 2000).

The expression below was used to convert the traces:

$$1 : \frac{Pp_{res}.\ \delta_{res}}{\delta_{cim}} : \frac{Pp_{cal}.\ \delta_{cal}}{\delta_{cim}} : \frac{Pp_{agr}.\ \delta_{agr}}{\delta_{cim}}$$

Where we can say that:

Pp_{res} = proportion of residue in the mortar mix, in apparent volume, in dm3;

Pp_{cal} = proportion of hydrated lime in the mortar mix, in apparent volume, in dm^3 ;

Pp_{agr} = proportion of fine aggregate in the mortar mix, in apparent volume, in dm3;

δ_{res} = unit mass of the waste, in its loose and dry state, in kg/dm3;

δ_{cal} = unit mass of hydrated lime, in the loose state, in kg/dm3;

δ_{agr} = unit mass of the fine aggregate, loose and dry, in kg/dm3;

δ_{cim} = unit mass of cement, in the loose state, in kg/dm3.

When converting the traits to be used from volume to mass, the following proportions were obtained between the materials (Table 5):

Volume trace	Mass trace without replacement*	Trace composition	Mass trace with substitution*	Trace composition	Nomenclature
1:1:6	333,30:210,30: 2.889,30 g	cement:lime:sand	316,60:18,00:210,40: 2.889,30 g	. cement:residue:lime:sand	A1/5/RPPC
			300,00:36,10:210,40: 2.889,30 g		A1/10/RPPC
			266,70:72,10:210,40: 2.889,30 g		A1/20/RPPC
			233,30:108,20:210,40: 2.889,30 g		A1/30/RPPC
1:1:6	333,30:210,30: 2.889,30 g	cement:lime:sand	316,70:15,90:210,40: 2.889,30 g	. cement:residue:lime:sand	A1/5/RPP
			300,00:31,90:210,40: 2.889,30 g		A1/10/RPP
			266,70:63,80:210,40: 2.889,30		A1/20/RPP
			233,30:95,70:210,40: 2.889,30 g		A1/30/RPP
1:0,5:4,5	434,80:137,20: 2.826,48 g	cement:lime:sand	413,00:23,50:137,20: 2.826,48 g	. cement:waste:lime:sand _	A2/5/RPPC
			391,30:47,00:137,20: 2.826,48 g		A2/10/RPPC
			347,80:94,00:137,20: 2.826,48 g		A2/20/RPPC
			304,50:141,09:137,20: 2.826,48 g		A2/30/RPPC
1:0,5:4,5	434,80:137,20: 2.826,48 g	cement:lime:sand	413,00:20,78:137,20: 2.826,48 g	. cement:residue:lime:sand _	A2/5/RPP
			391,30:41,61:137,20: 2.826,48 g		A2/10/RPP
			347,80:83,22:137,20: 2.826,48 g		A2/20/RPP
			304,50:124,83:137,20: 2.826,48 g		A2/30/RPP

*replacement of different levels of cement with limestone quarry waste or porcelain polishing waste.

Table 5 - Mortar mix conversion.

4.6 MORTAR PREPARATION

The materials used in the research were weighed on duly calibrated electronic scales. The cement, waste, lime and water were weighed using a scale with a load capacity of 1000 g and a resolution of 0.01 g, brand HELMAC, model HM1000. The fine aggregate was weighed on an electronic scale with a load capacity of 10,000 g and a resolution of 0.1 g, brand MARTE, model AC 10K.

In this experiment, the mortars were produced using a PAVITEST C 3010-146-380 5-liter vertical shaft mechanical mixer (mortar mixer), as shown in Figure 16.

Source: author.

Figure 16 - Mechanical mixer.

First, the water content was determined to obtain the standard consistency index prescribed in NBR 13276 (ABNT, 2005), using the reference mortars (without residue). Having reached a consistency in the 260 + 5 mm range recommended by NBR 13276 (ABNT, 2005), the water/cement ratio was set for each of the reference mixes, as shown in Table 6, where the values found were also used in the mixes containing residues.

In order to achieve greater control over production, we decided to separate all the material to be used in making the trace beforehand.

Mortar	fa/c	$f_{a/agl}$	fa/mat.seco	fc/mat$_{.seco}$	$f_{c/cal}$	CI (mm)
A1	1,87	1,15	0,18	0,10	1,58	258,00
A2	1,45	1,10	0,19	0,13	3,17	261,00

Table 6 - Water/cement ratio (a/c), water/binder (a/agl.), water/dry materials (a/dry materials), cement/dry materials (c/dry materials), cement/cal (c/cal) and consistency index (CI).

4.7 TEST METHODS FOR CHARACTERIZING MORTARS

4.7.1 Fresh mortar

4.7.1.1   consistency index - workability

Standard NBR 13276 (ABNT, 2005) was used in this test as it prescribes the method for determining the water content in the preparation of mortars for laying walls and covering walls and ceilings. The principle of the test is to measure the spread of the mortar after it has been subjected to 30 blows on the consistency table.

To carry out the test, a flow table was used, as described in NBR 7215 (ABNT, 1996). The principle of the test is to measure the spread of the mortar after it has been subjected to 30 blows on the consistency table. Figures 17a, b and c illustrate the sequential stages of the test.

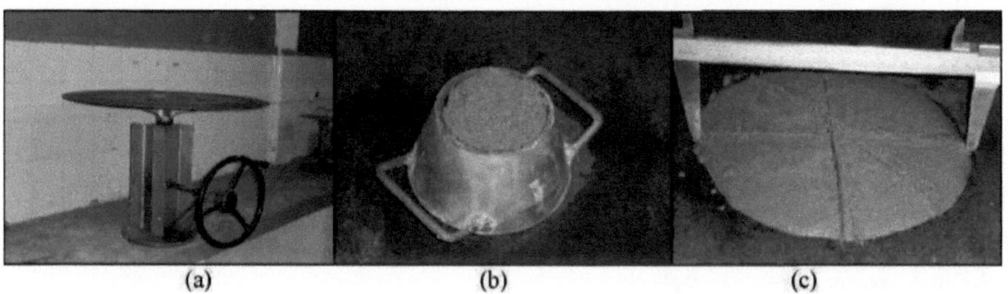

Source: author.

Figure 17a - "Flow table".

Figure 17b - Cone trunk used in the test.

Figure 17c - Measuring the spread of the mortar.

4.7.1.2   Water retention

With regard to the water retention test for mortars, at the time the test was carried out, the laboratory did not have a Buchner funnel so that the test could be carried out as prescribed by NBR 13277 (ABNT, 2005). For this reason, the instructions contained in NBR 13277 (ABNT, 1995) were used.

Although authors such as Nakakura (2004) and Tristao and Machado (2003), among others, have found that the NBR 13277 method (ABNT, 1995) is not effective, it was assumed that the results found from the test would only be comparative between the mortar types produced in the research, regardless of the method used.

The basic principle of this test is to measure the water retention capacity of mortars. Figure 18 illustrates the equipment needed to carry out the test.

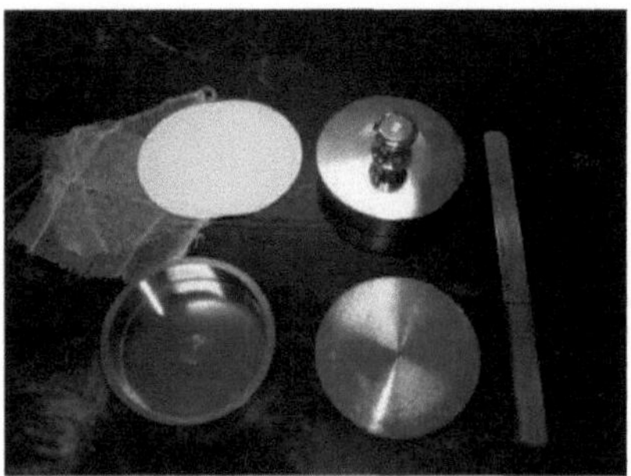

Source: author.

Figure 18 - Test apparatus.

4.7.1.3 Mass density and incorporated air content

Mass density and incorporated air content were determined using the procedures established in NBR 13278 (ABNT, 2005). This standard describes the method for determining the density of mass and the content of incorporated air in mortars in the fresh state, intended for laying and coating walls and ceilings. The test result shows the state of aggregation of the particles present in the mixture.

4.7.2 Mortar in the hardened state

To carry out the tests with the mortars in the hardened state, when the specimens were finished being molded, the upper side of the mold was protected by a flat glass plate measuring 7 x 7 cm and 4 mm thick, so that the water would not evaporate during the mortar hardening process. After 48 ± 24 hours, the specimens were demolded and identified by batch with the date of molding, the mix used and the sequence number. All the specimens were subjected to air curing in a covered area at the CEFET/RN Civil Construction Materials Laboratory for a minimum of 28 days before being tested.

4.7.2.1   Apparent mass density in the hardened state

The test for apparent mass density in the hardened state was carried out at the age of 28 days, in accordance with the requirements set out in NBR 13280 (ABNT, 2005). This prescribes the test method for determining the bulk density of mortar for laying and coating walls and ceilings, in the hardened state. The test specimens were molded in accordance with the requirements of NBR 13279 (ABNT, 2005).

4.7.2.2   Modulus of elasticity

The modulus of elasticity test was carried out in accordance with the requirements of draft standard no. 18:400.04-008 (ABNT, 2008), entitled "Mortar for wall and ceiling laying and coating -

Determination of the dynamic modulus of elasticity by means of ultrasonic wave propagation". The specimens were molded in accordance with NBR 13279 (ABNT, 2005) and tested at 28 days of age. The results of the tests show the mortar's capacity to deform when subjected to stress.

The test was carried out using a CONTROLS ultrasonic pulse velocity meter, model 58 - E0048. Figure 19 shows the equipment used and how the test was carried out.

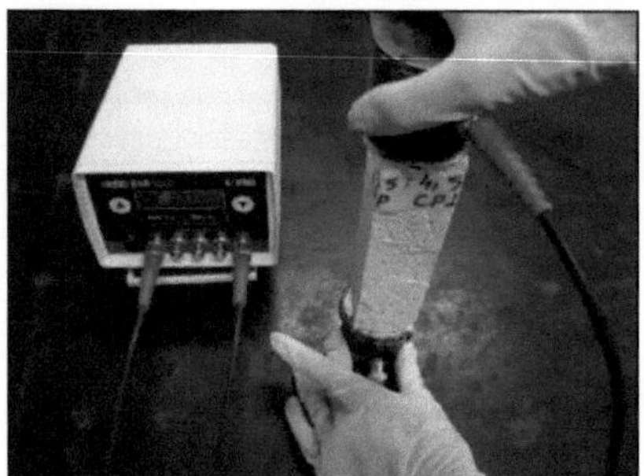

Source: author.

Figure 19 - Equipment used in the test.

4.7.2.3 Tensile strength in bending

The flexural tensile strength was determined at the age of 28 days, according to the procedure described in NBR 13279 (ABNT, 2005).

When the specimens were molded, the drops were applied through the table to determine the consistency index, as the laboratory did not have the densification table prescribed by the standard. It should be noted that all the specimens were molded equally.

Before the flexural tensile strength test was carried out, the axis of the specimen was marked to ensure that the load was applied in the center of the specimen. The test was carried out in the mechanical test room of the UFRN Cement Laboratory on a Shimadzu universal press model Autograph AG-I - TRAPEZIUM 2. Figure 20 illustrates how the test was carried out.

Source: author.

Figure 20 - Flexural tensile strength test.

4.7.2.4 Water absorption by capillarity

The test method described in NBR 15259 (ABNT, 2005) was followed for this test. The specimens were molded in accordance with NBR 13279 (ABNT, 2005) and tested at 28 days of age. The basic principle of this test is to measure the absorption of water in mortars through capillary pores. Figure 21 shows when the specimens were tested.

Source: author.

Figure 21 - Specimens in contact with water.

4.7.2.5   Microstructural analysis

Progress in the field of materials has mainly resulted from the recognition of the principle that a material's properties originate from its internal structure; in other words, properties can be modified by appropriate changes in the material's structure (MEHTA and MONTEIRO, 1994). In view of this, the formulated mortars were analyzed microstructurally in order to help find answers when characterizing them.

The samples for the microstructural analysis were taken from cylindrical specimens measuring 5 x 10 cm, molded exclusively for the test. The specimens were molded in two layers of approximately equal heights, with each layer receiving 10 blows applied through the table to determine the consistency index.

For microscopic visualization, it is necessary that the observed surface has not received any type of friction or any physical incidence that could alter the surface tensions of the material, so as not to modify the intrinsic characteristics of the observed mortar (SOUZA, 2007).

At ages over 28 days, the specimens were manually fractured by impact using a metal bar, where they were reduced into smaller samples measuring approximately 0.70 x 0.50 x 0.50 cm and stored in plastic containers with lids. The samples were then metallized with gold to increase visual contrast.

The micrographs were taken at the Institutional Scanning Electron Microscopy Laboratory - LIMEV - UFRN, using a Phillips model XL 30 - ESEM.

The same analysis criteria were used for all the samples, where the best metallized point was selected first, for better visualization, and then it was magnified, always starting from the central region, by 3000 and 5000 times.

To analyze the elemental chemical composition of the samples, some of them were selected because they were representative and had very different concentrations of residues. The analysis was carried out using energy dispersive X-ray spectroscopy (EDS) - coupled with SEM. Table 7 shows the mortars analyzed.

Specimen identification	Identification of the sample analyzed
1:1:60%	1
1:1:630%RPPC	3
1:1:630%RPP	8
1:0,5:4,50%	4
1:0.5:4.530%RPPC	6
1:0.5:4.530%RPP	10

Table 7 - Samples selected for microscopic and EDS analysis.

4.7.2.6   XRD and XRF analysis

X-ray diffraction (XRD) and X-ray fluorescence (XRF) techniques were used to identify the phases formed in all the mortars produced. The samples for these tests were extracted from the same specimens used to obtain the samples for microstructural analysis. A small sample of the specimen was crushed and ground in a mortar and the fraction passing through a #50 sieve was used. The mortar types analyzed by X-ray diffraction and X-ray fluorescence were the same as those previously selected for microscopic analysis (Table 7).

Phase analysis by X-ray diffraction was carried out using Shimadzu XRD-7000 equipment, located in the Cement Laboratory at UFRN. The X-ray diffraction experiments were carried out using Cu-Kα radiation, with an accelerating voltage of 30 kV and a current of 30 mA, scanning from 5° to 80°, at a scanning speed of 2°/min. In the case of X-ray fluorescence analysis, use was made of the Shimadzu EDX 720 equipment, which is located in the Materials Characterization Laboratory at UFRN.

CHAPTER 5

RESULTS AND DISCUSSION

5.1 CHARACTERIZATION OF STARTING MATERIALS

5.1.1 Specific mass and unit mass

Table 8 shows the results of the specific masses and unit masses of the materials used in the experimental study.

Essay	Method	Results
Cement		
Specific mass (g/cm3)	NBR 9776 (1987)	2,98
Unit mass (g/cm)^3	NBR 7251 (1982)	1,06
Cal		
Specific mass (g/cm3)	NBR 9776 (1987)	2,70
Unit mass (g/cm3)	NBR 7251 (1982)	0,65
Small aggregate (sand)		
Specific mass (g/cm3)	NBR 9776 (1987)	2,62
Unit mass (g/cm3)	NBR 7251 (1982)	1,54
Limestone quarry waste		
Specific mass (g/cm3)	NBR 9776 (1987)	2,50
Unit mass (g/cm3)	NBR 7251 (1982)	1,47
Porcelain tile polishing residue		
Specific mass (g/cm3)	NBR 9776 (1987)	2,25
Unit mass (g/cm3)	NBR 7251 (1982)	1,01

Table 8 - Specific mass of the starting materials.

The results show that the specific masses of the waste are lower than the specific masses of the ground aggregate and the binders used in the experiment. Therefore, the waste has a lower specific surface than the other materials. The specific surface of the material is directly proportional to the bond between the aggregate and the binders, and therefore to the aggregation of the particles. Furthermore, Nakakura (2000) apud SILVA et al (2005) points out that the specific mass is directly proportional to the fines content of the material. Therefore, as RPPC has a higher specific mass than RPP, it has a higher fines content in its composition.

In terms of unit mass, the results show that the residues have a higher unit mass than the binders and lower than the fine aggregate. According to Pandolfo et al (2005), unit mass is inversely proportional to the compactness of mortars. Residues can therefore help to fill the voids in the aggregate/binder bond.

5.1.2 Granulometry

5.1.2.1　Small aggregate

The fines content and particle size are decisive elements for the performance of the mortars. Table 9 shows the particle size characteristics of the fine aggregates used in the mortars studied. Figure 22 illustrates the grain size curve of the sand used in the mixtures.

Sieve mesh opening (mm)	% retained by mass	% retained in accumulated mass	Method
4,8	0,00	0,00	
2,4	12,768	63,84	
1,2	18,612	93,06	
0,6	17,486	87,43	NBR 7217
0,3	16,850	84,25	(ABNT, 1987)
0,15	13,426	67,13	
< 0,15	20,858	104,29	
Total	100,000	-	
Fineness modulus		2,38	NBR 7217 (ABNT, 1987)
Maximum characteristic dimension (mm)		4,8	NBR 7217 (ABNT, 1987)
Graduation (zone)	Zone 3 (medium sand)		NBR 7211 (ABNT, 1983)

Table 9 - Granulometry of the fine aggregate.

Error! Link not valid.

Figure 22 - Sand particle size curve.

Table 9 shows that the sand used meets the appropriate standards for use in the designs studied, as it has a particle size gradation as established by NBR 7211 (ABNT, 1983) and a maximum characteristic size (mm) as prescribed by NBR 7217 (ABNT, 1987). Figure 22 shows the grain size curve of the sand in the interval between the lower and upper bands of zone 3, classifying it as medium sand.

5.1.2.2　Limestone quarry waste

Laser granulometry was carried out on the waste used in this research. Table 10 shows the results of the test.

Limestone quarry waste	D10%	D50%	D90%

Limestone quarry waste	D10%	D50%	D90%
	0,82	20,56	194,36
Average particle diameter	65,60		

Table 10 - Laser granulometry of RPPC (diameter in µm).

The test results show that the average particle size of limestone quarry waste is 65.60 µm. Anhydrous Portland cement consists of angular particles commonly between 1 and 50 µm in size (MEHTA and MONTEIRO, 1994). In view of this, it can be concluded from the laser granulometry that the limestone quarry waste probably does not have an adequate granulometry to fill the voids left by the cement grains.

According to Silva (2005), in terms of particle size, the system can be classified as:

-Granular system : particles over 44 µm, gravitational forces predominate, material flows freely.

-Colloidal system : particles smaller than 1 µm, surface forces predominate, controlling the behavior of the system.

-Dust: particles from 1 to 44 µm, gravitational and surface forces of the same order.

According to the above classification, it can be seen that limestone quarry waste behaves like a granular system.

Figure 23 illustrates the particle diameter as a function of the accumulated percentage of limestone rock waste (RPPC).

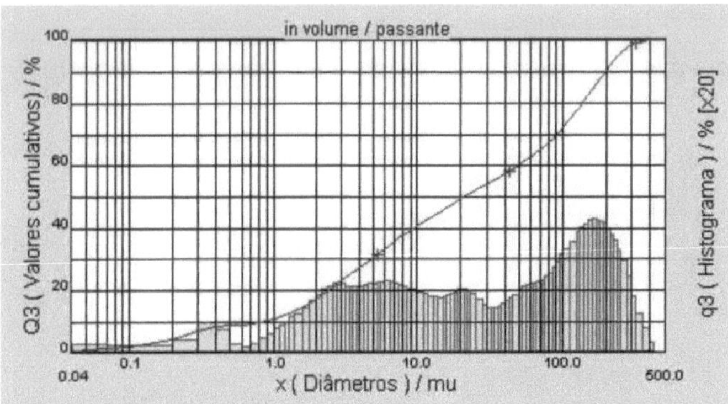

Figure 23 - Particle size distribution of RPPC.

5.1.2.3   Residue from polishing porcelain tiles

Table 11 shows the results of the laser granulometry of the porcelain tile polishing residue. Figure 24 illustrates the particle size curve found for this residue.

Porcelain polish residue	D10%	D50%	D90%
	1,60	13,34	52,08
Average particle diameter	20,85		

Table 11 - Laser granulometry of RPP (diameter in μm).

Based on the particle size, it can be said that the porcelain tile polishing residue is possibly a suitable material for filling the voids left by the cement grains, since 80% of the particles have a diameter of 20.85 μm. In terms of classification, it behaves like a powder, as the diameter of its particles is between 1 and 44 μm.

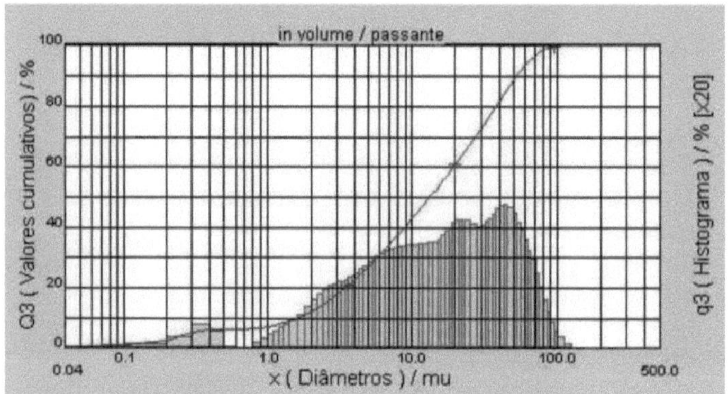

Figure 24 - Particle diameter x Cumulative percentage of RPP.

5.1.3 Scanning electron microscopy

5.1.3.1   Limestone quarry waste

Scanning electron microscopy showed the morphology of the waste particles, as illustrated in Figure 25.

Figure 25 - Micrograph of limestone quarry waste - back scattered electrons.

The micrograph shows that the particles of the limestone quarry waste have an irregular and angular shape. The demand for water is generally related to the shape of the particles, thus characterizing the rolling effect for the spherical particles which favours the workability of the mortar, unlike the irregular and angular particles of the limestone quarry waste.

5.1.3.2　　Residue from polishing porcelain tiles

Figure 26 shows a scanning electron microscopy image of the porcelain tile polishing residue. The analysis shows the morphology of the residue particles. The micrographs show that the particles of the porcelain tile polishing residue are even more irregular and angular than those of the limestone quarry residue. As a result, the workability of mortars dosed with this material tends to be lower.

Figure 26 - Micrograph of porcelain tile polishing residue - secondary electrons.

5.1.4 X-ray diffractometry and X-ray fluorescence

5.1.4.1　　Limestone quarry waste

Figure 27 shows the X-ray diffractogram of limestone quarry waste. The results of the analysis show that there is a predominance of the $CaCO_3$, CaO and SiO_2 phases. These phases were possibly formed due to the limestone origin of the waste, as well as the presence of quartz.

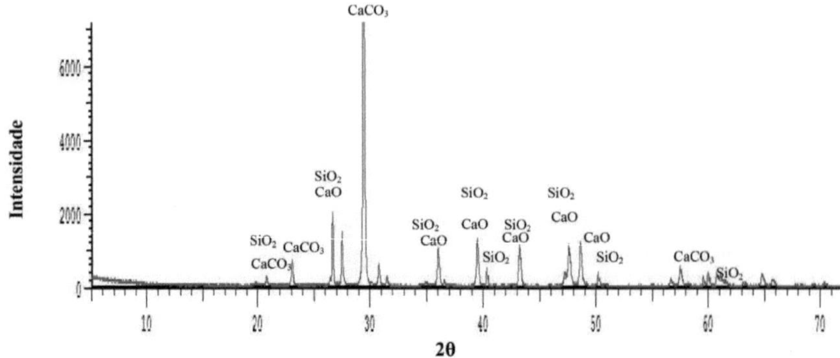

Figure 27 - XRD of RPPC.

Table 12 shows the results of the X-ray fluorescence analysis. The results show a predominance of Ca and Si, confirming the results of the X-ray diffraction analysis. In addition, Al and Fe were found, probably associated with phases present at lower levels.

Phase	Percentage
CaO	77,568%
SiO2	11,866%
Al2O3	5,261%
Fe2O3	2,231%
SO3	1,401%
K2O	1,251%
Others	0,422%

Table 12 - X-ray fluorescence of RPPC.

5.1.4.2 Residue from polishing porcelain tiles

X-ray diffractometry of the porcelain tile polishing residue, as shown in Figure 28, shows that the predominant phases are quartz, mullite and silicon carbide. The presence of quartz and mullite possibly comes from the porcelain tile, while the presence of silicon carbide may be related to the abrasive used in the porcelain tile polishing process.

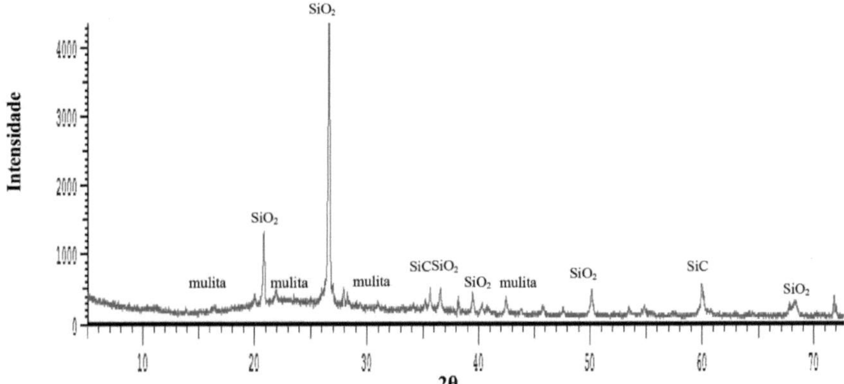

Figure 28 - XRD of RPP.

Table 13 shows the X-ray fluorescence analysis of RPP. The analysis shows Si as the predominant element, thus confirming the formation of the quartz found in the X-ray diffraction.

Phase	Percentage
SiO_2	59,433%
Al2O3	18,989%
MgO	10,953%
K2O	3,050%
ZrO2	2,260%
CaO	1,911%
SO3	1,553%
P2O5	1,010%
Others	0,841%

Table 13 - X-ray fluorescence of RPP.

5.2 FRESH MORTAR TEST RESULTS

5.2.1 consistency index - workability

Table 14 shows the average results of the consistency indices obtained in all the tests carried out with the A1 and A2 mortars. Table 15 shows the percentage variation in these indices for all the formulated mixes.

Mortar	A/c ratio	consistency index (mm)	Standard deviation	Coefficient of variation (%)
A1/0/REF		258	4,89	1,89
A1/5/RPPC		235	4,05	1,72
A1/10/RPPC		247	4,55	1,84

Mortar	f_{a/agl}			
A1/20/RPPC		240	5,88	2,45
A1/30/RPPC	1,87	236	0,95	0,40
A1/5/RPP		208	5,97	2,87
A1/10/RPP		221	4,39	1,98
A1/20/RPP		211	4,41	2,09
A1/30/RPP		211	2,78	1,32
A2/0/REF		261	5,57	2,13
A2/5/RPPC		259	4,05	1,56
A2/10/RPPC		248	6,05	2,44
A2/20/RPPC		257	8,74	3,40
A2/30/RPPC	1,45	244	7,95	3,26
A2/5/RPP		252	5,82	2,31
A2/10/RPP		241	4,55	1,89
A2/20/RPP		245	6,11	2,49
A2/30/RPP		243	1,65	0,68

Table 14 - Average results of the mortar consistency indices.

Mortar	$f_{a/agl}$	$F_{a/res}$	$f_{cal/cim}$	Variation between the consistency of mortars with residues and reference mortars (%)
A1/5/RPPC	1,18	34,67	0,66	-8,91
A1/10/RPPC	1,22	17,29	0,70	-4,26
A1/20/RPPC	1,31	8,65	0,79	-6,98
A1/30/RPPC	1,41	5,77	0,90	-8,53
A2/5/RPPC	1,15	26,81	0,33	-0,77
A2/10/RPPC	1,19	13,40	0,35	-4,98
A2/20/RPPC	1,30	6,7	0,39	-1,53
A2/30/RPPC	1,43	4,47	0,45	-6,51
A1/5/RPP	1,18	39,25	0,66	-19,38
A1/10/RPP	1,22	19,56	0,70	-14,34
A1/20/RPP	1,31	9,78	0,79	-18,22
A1/30/RPP	1,41	6,52	0,90	-18,22
A2/5/RPP	1,15	30,32	0,33	-3,45
A2/10/RPP	1,19	15,14	0,35	-7,66
A2/20/RPP	1,30	7,57	0,39	-6,13
A2/30/RPP	1,43	5,05	0,45	-6,90

Table 15 - Variation of the consistency index between mortars with residues and reference mortars.

In this test, it was found that there was a decrease in the consistency index of the mortars produced with waste and, as a consequence, a probable increase in cohesiveness, regardless of the mix.

It was also observed that as the lime/cement ratio increases, the need for water to obtain the standard consistency index increases. This is probably due to the increase in the specific surface area of the mortar's binder fraction, since lime is thinner than cement.

In the case of mortars A1 and A2 dosed with RPPC and A2 made with RPP, the decrease in the consistency index was less than 9%. For the A1 mortars formulated with RPP, the decrease in consistency varied between 14% and 20%. In this case, it can be said that the RPP is probably contributing to a better bond between the aggregate and the binder paste, thus acting as an agent capable of increasing the mortar's cohesion. Such an increase in cohesion can provide gains in terms of quality and productivity when raising and coating masonry, and can also influence the reduction of mortar losses, thus reducing waste and, consequently, waste generation.

When comparing the consistency indices of mortars with the same design in relation to the type of waste used in their formulation, it can be seen that the mortars produced with RPPC (specific mass of 2.50 g/cm^3) had a lower consistency index than those produced with RPP (specific mass 2.25 g/cm3). In view of this, it is likely that the higher the specific mass of the material, the less water is needed for it to reach the desired consistency index. In addition, microscopic analysis of the residues showed that RPP has a more angular particle morphology than RPPC, thus causing a more pronounced decrease in the workability of the mortars. It should also be noted that when cement is replaced by waste, there is a reduction in the amount of binder in the mix, since cement is finer than waste.

Figure 29 shows that there is a correlation between the consistency indices and the percentages and types of waste used in the mortars. However, the correlation or dependence shown is probabilistic. The trend lines indicate that the presence of residues in the mix can lead to a reduction in the consistency index of the mortars as the residue content increases.

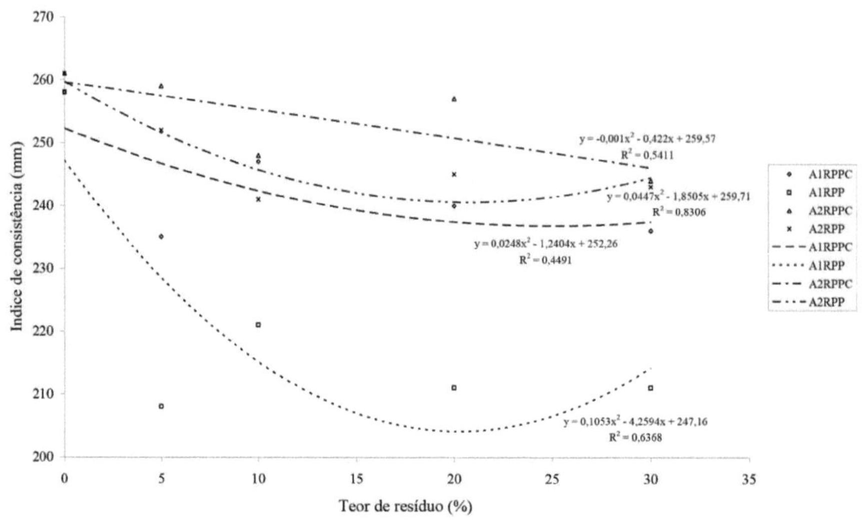

Figure 29 - Correlation between residue content and mortar consistency index.

5.2.2 Water retention

Table 16 shows the average results of the water retention test for the mortars produced. Table 17 shows the percentage variation in water retention between the mortars with residues and the reference mortars.

Mortar	Retention (%)
A1/0/REF	94,33
A1/5/RPPC	98,25
A1/10/RPPC	95,49
A1/20/RPPC	97,41
A1/30/RPPC	99,02
A1/5/RPP	96,28
A1/10/RPP	93,95
A1/20/RPP	96,05
A1/30/RPP	96,60
A2/0/REF	90,85
A2/5/RPPC	92,29
A2/10/RPPC	92,61
A2/20/RPPC	93,30
A2/30/RPPC	92,66
A2/5/RPP	90,47

Mortar	
A2/10/RPP	92,59
A2/20/RPP	91,80
A2/30/RPP	96,27

Table 16 - Average water retention results for mortars.

Mortar	Variation between water retention of mortars with residues and reference mortars (%)
A1/5/RPPC	+4,16
A1/10/RPPC	+1,23
A1/20/RPPC	+3,27
A1/30/RPPC	+4,97
A1/5/RPP	+2,07
A1/10/RPP	-0,40
A1/20/RPP	+1,82
A1/30/RPP	+2,41
A2/5/RPPC	+1,59
A2/10/RPPC	+1,94
A2/20/RPPC	+2,70
A2/30/RPPC	+1,99
A2/5/RPP	-0,42
A2/10/RPP	+1,92
A2/20/RPP	+1,05
A2/30/RPP	+5,97

Table 17 - Variation in water retention between mortars with residues and reference mortars.

The results show that there was no significant change, i.e. increase or decrease, in the water retention of the mortars formulated, regardless of the percentage or type of residue used in the mix. In view of this, the use of residues in the mortar mix as a substitute for cement probably does not influence this property.

However, when correlating the percentage of residue present in the mix with the percentage of water retained (Figure 30), it can be seen that all the mortars produced with residue tend to increase water retention as the residue content increases. This could possibly be explained by the fact that the residues are replacing cement. Therefore, the lime/cement ratio will be higher as the percentage of residue in the mixture increases, since lime is thinner than cement and therefore tends to retain more water.

It should be noted that the A2 RPP mix, starting at 20%, tends to reduce water retention. Consequently, higher contents will lead to a reduction in mortar strength, as cement hydration and lime carbonation will be impaired due to inadequate water loss. This can be explained not only by the fact that the A2 mix has a lower lime/cement ratio than the A1 mix, but also by the fact that the RPP has a lower specific mass than the RPPC.

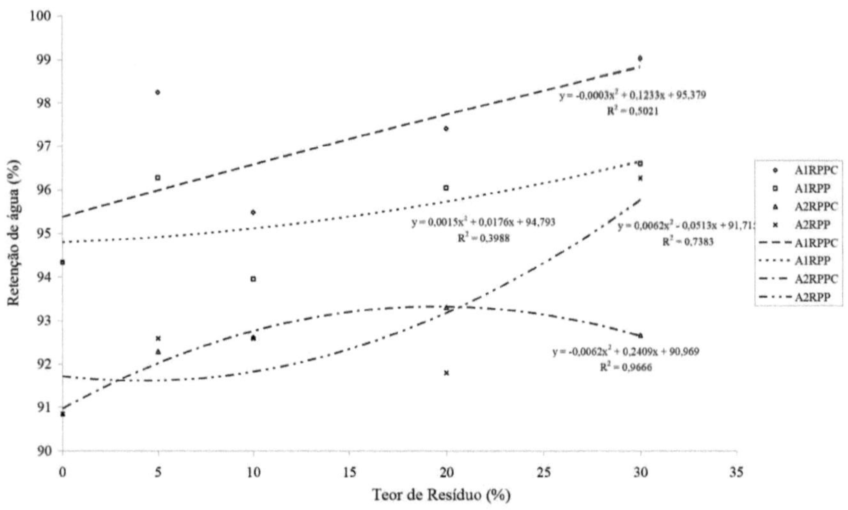

Figure 30 - Correlation between residue content and water retention of mortars.

With regard to the classification of mortars, NBR 13281 (ABNT, 2005), whose classification can be found in Annex C, prescribes that for mortars to have good performance and durability, they must have adequate water retention. Consequently, they must be classified in the higher bands of the classification table.

In Table 18, the results show that the mortars studied comply with NBR 13281:2005, since they were classified in the higher ranges of the table, as determined by the standard, and that there is no relevant difference between the mortars formulated with residue and the reference mortars.

Mortar	Retention (%)	Classification - NBR 13281:2005
A1/0/REF	94,33	U5
A1/5/RPPC	98,25	U6
A1/10/RPPC	95,49	U5
A1/20/RPPC	97,41	U6
A1/30/RPPC	99,02	U6
A1/5/RPP	96,28	U6
A1/10/RPP	93,95	U5
A1/20/RPP	96,05	U6
A1/30/RPP	96,60	U6

A2/0/REF	90,85	U4
A2/5/RPPC	92,29	U5
A2/10/RPPC	92,61	U5
A2/20/RPPC	93,30	U5
A2/30/RPPC	92,66	U5
A2/5/RPP	90,47	U4
A2/10/RPP	92,59	U5
A2/20/RPP	91,80	U5
A2/30/RPP	96,27	U5

Table 18 - Classification of mortars according to NBR 13281:2005.

5.2.3 Mass density and incorporated air content

The results of this test are shown in Tables 19 and 20 and Figures 31 and 32. The results relating to mass density will be discussed first, followed by those relating to incorporated air content.

5.2.3.1 Mass density

Tables 19 and 20 show the results of the mass density test.

Mortar	Mass density $(Kg/m)^3$
A1/0/REF	1545
A1/5/RPPC	1522
A1/10/RPPC	1510
A1/20/RPPC	1554
A1/30/RPPC	1538
A1/5/RPP	1546
A1/10/RPP	1546
A1/20/RPP	1565
A1/30/RPP	1531
A2/0/REF	1589
A2/5/RPPC	1575
A2/10/RPPC	1576
A2/20/RPPC	1575
A2/30/RPPC	1573
A2/5/RPP	1582
A2/10/RPP	1585
A2/20/RPP	1594
A2/30/RPP	1580

Table 19 - Average mass density results for mortars.

Mortar	Variation between the mass density of mortars with residues and reference mortars (%)
A1/5/RPPC	-1,49
A1/10/RPPC	-2,27
A1/20/RPPC	0,58
A1/30/RPPC	-0,45
A2/5/RPPC	-0,88
A2/10/RPPC	-0,82
A2/20/RPPC	-0,88
A2/30/RPPC	-1,01
A1/5/RPP	+0,06
A1/10/RPP	0,00
A1/20/RPP	+1,29
A1/30/RPP	-0,91
A2/5/RPP	-0,44
A2/10/RPP	-0,25
A2/20/RPP	+0,31
A2/30/RPP	-0,57

Table 20 - Variation in mass density between mortars with residues and reference mortars.

Density corresponds to the state of aggregation of the particles of the mortar's component materials. The results shown in Tables 19 and 20 show that the presence of waste in the production of mortars, regardless of the percentage or type, probably does not alter the mass density of mortars in the fresh state.

However, for the mortars produced with RPPC, when correlating the percentage of residue present in the mixtures with the bulk density (Figure 31), we can see a tendency for the density to increase. This is probably due to the fact that RPPC has a higher specific mass, which means greater aggregation and packing of the particles. It should also be noted that the morphology of RPPC particles is less angular than that of RPP. As a result, the mortars produced with RPP tended to reduce their bulk density when the residue was present in the mixtures, regardless of the mix design.

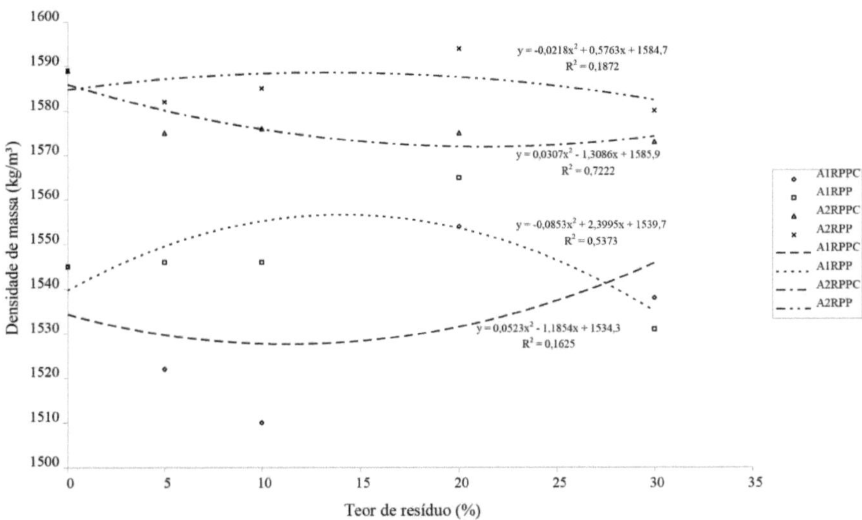

Figure 31 - Correlation between residue content and mass density in the fresh state of the mortars.

The classification of the mortars studied according to NBR 13281 (ABNT, 2005), whose classification can be found in Appendix C, is shown in Table 21. This standard recommends that mortars have a mass density in the lower ranges of the classification so that their performance and durability are not compromised. For the formulated mortars, the results show that there is no variation between the mortars formulated with residue and the reference mortar and that the mortars are in the central range of the classification table established by the standard.

Mortar	Mass density (Kg/m^3)	Classification - NBR 13281:2005
A1/0/REF	1545	D3
A1/5/RPPC	1522	D3
A1/10/RPPC	1510	D3
A1/20/RPPC	1554	D3
A1/30/RPPC	1538	D3
A1/5/RPP	1546	D3
A1/10/RPP	1546	D3
A1/20/RPP	1565	D3
A1/30/RPP	1531	D3
A2/0/REF	1589	D3
A2/5/RPPC	1575	D3
A2/10/RPPC	1576	D3

Mortar		
A2/20/RPPC	1575	D3
A2/30/RPPC	1573	D3
A2/5/RPP	1582	D3
A2/10/RPP	1585	D3
A2/20/RPP	1594	D3
A2/30/RPP	1580	D3

Table 21 - Classification of mortars according to NBR 13281:2005.

5.2.3.2 Incorporated air content

Table 22 shows the results of the incorporated air content test on the formulated mortars. Table 23 shows the percentage variation between the mortars with residue and the reference mortars.

Mortar	Incorporated air content (%)
A1/0/REF	41
A1/5/RPPC	42
A1/10/RPPC	42
A1/20/RPPC	40
A1/30/RPPC	41
A1/5/RPP	41
A1/10/RPP	41
A1/20/RPP	40
A1/30/RPP	41
A2/0/REF	40
A2/5/RPPC	40
A2/10/RPPC	40
A2/20/RPPC	40
A2/30/RPPC	40
A2/5/RPP	40
A2/10/RPP	40
A2/20/RPP	39
A2/30/RPP	40

Table 22 - Average results of the incorporated air content of the mortars.

Mortar	Variation between the incorporated air content of mortars with residues and reference mortars (%)
A1/5/RPPC	2,44
A1/10/RPPC	2,44
A1/20/RPPC	-2,44
A1/30/RPPC	0,00

A2/5/RPPC	0,00
A2/10/RPPC	0,00
A2/20/RPPC	0,00
A2/30/RPPC	0,00
A1/5/RPP	0,00
A1/10/RPP	0,00
A1/20/RPP	-2,44
A1/30/RPP	0,00
A2/5/RPP	0,00
A2/10/RPP	0,00
A2/20/RPP	-2,50
A2/30/RPP	0,00

Table 23 - Variation in incorporated air content between mortars with residues and reference mortars.

Generally, the increase in the content of incorporated air occurs during the mixing operations of the materials in the preparation of the mortars and in the molding of the specimens for testing. As there was standardization in the preparation and moulding procedures, the presence of residues in the mix probably did not alter this property.

However, the correlation (Figure 32) between the mixes studied showed that there was a tendency for the air content of mortars produced with RPP to increase when compared to those produced with RPPC. This is possibly due to the fact that RPP has a lower specific mass than RPPC, thus causing more air to be trapped in the mixtures.

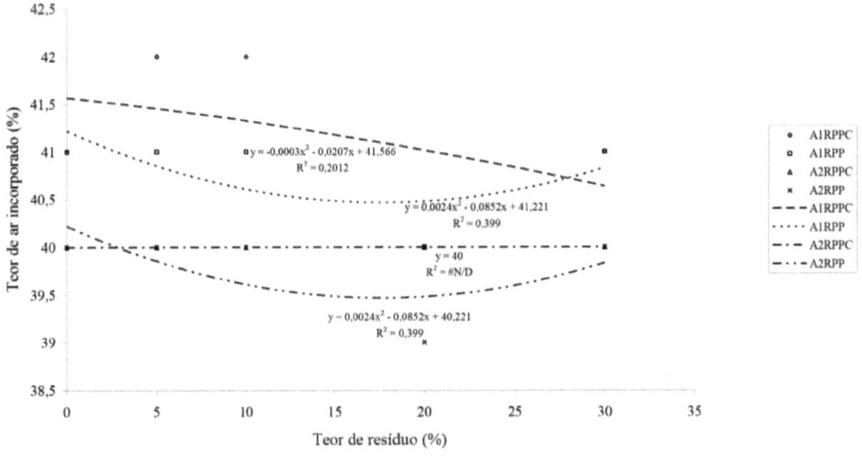

Figure 32 - Correlation between residue content and incorporated air content of mortars.

5.3 RESULTS OF THE MORTAR TESTS IN THE HARDENED STATE

5.3.1 Apparent mass density

Table 24 shows the average results of the bulk density tests carried out. Table 25 shows the variation in bulk density between the mortars with residues and the reference mortars.

Mortar	Mass density apparent (kg/m>)	Coefficient of Standard deviation	variation (%)
A1/0/REF	1750	9,88	0,56
A1/5/RPPC	1771	6,01	0,34
A1/10/RPPC	1743	15,33	0,88
A1/20/RPPC	1694	1,37	0,08
A1/30/RPPC	1729	4,79	0,28
A1/5/RPP	1751	7,12	0,41
A1/10/RPP	1767	12,17	0,69
A1/20/RPP	1758	15,91	0,91
A1/30/RPP	1746	3,49	0,20
A2/0/REF	1806	10,92	0,60
A2/5/RPPC	1789	13,81	0,77
A2/10/RPPC	1766	8,54	0,48
A2/20/RPPC	1254	11,44	0,64
A2/30/RPPC	1730	4,08	0,24
A2/5/RPP	1741	3,67	0,21
A2/10/RPP	1749	2,58	0,15
A2/20/RPP	1733	10,51	0,61
A2/30/RPP	1720	20,21	1,17

Table 24 - Average results of mortar bulk density.

Mortar	Variation between the bulk density of mortars with residues and reference mortars (%)
A1/5/RPPC	+1,20
A1/10/RPPC	-0,40
A1/20/RPPC	-3,20
A1/30/RPPC	-1,20
A2/5/RPPC	-0,94
A2/10/RPPC	-2,21
A2/20/RPPC	-30,56*
A2/30/RPPC	-4,21
A1/5/RPP	+0,06
A1/10/RPP	+0,97
A1/20/RPP	+0,46

A1/30/RPP	-0,23
A2/5/RPP	-3,60
A2/10/RPP	-3,16
A2/20/RPP	-4,04
A2/30/RPP	-4,76

*The value should be disregarded. In this case, there must have been a fault when the test was carried out.

Table 25 - Variation in bulk density between mortars with residue and reference mortars.

It was found that the presence of residues in the mixture causes a variation of less than 5% in this property.

According to Nakakura (2004), the bulk density value is an indication of the compactness resulting from the aggregate/binder mix ratio and the particle size distribution of the mix. It indirectly determines the volume of voids incorporated by the additives and the amount of water lost through evaporation. Therefore, the presence of residues in the mix as a substitute for cement probably does not compromise the good performance of modified mortars.

However, the correlations found between the traits studied show that the mortars produced with waste tend to reduce their mass density as the waste content in the mixture increases (Figure 33). This may have been due to the reduction in the specific mass of the mixture when cement was replaced by waste.

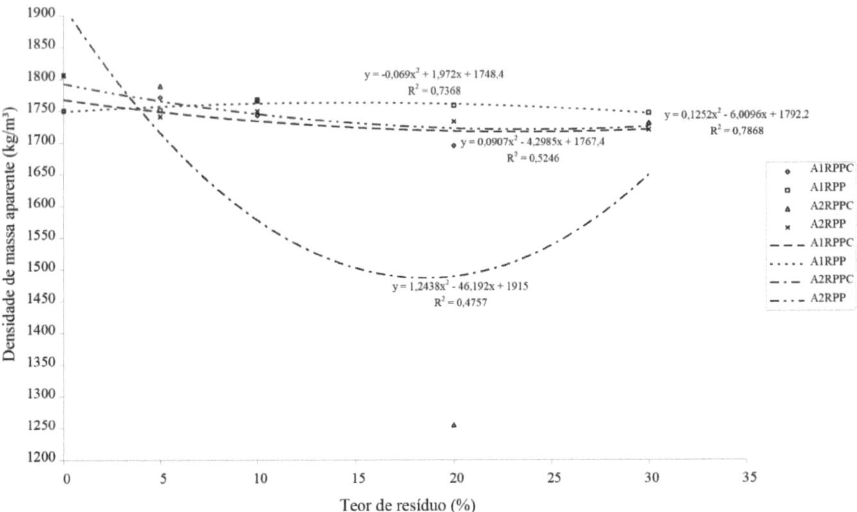

Figure 33 - Correlation of the residue content with the apparent mass density of the mortars.

In terms of classification, according to NBR 13281 (ABNT, 2005), Annex C, mortars must have a density in the lower ranges of the table. Table 26 shows the results of the mortars studied. It can be seen that there is no difference in classification between the mortars formulated with waste and the reference mortar. It can also be seen that the mortars are in the high range of the classification table. As a result, their performance and durability may be compromised when they are applied.

Mortar	Apparent mass density (kg/m)^3	Classification - NBR 13281:2005
A1/0/REF	1750	M5
A1/5/RPPC	1771	M5
A1/10/RPPC	1743	M5
A1/20/RPPC	1694	M5
A1/30/RPPC	1729	M5
A1/5/RPP	1751	M5
A1/10/RPP	1767	M5
A1/20/RPP	1758	M5
A1/30/RPP	1746	M5
A2/0/REF	1806	M5
A2/5/RPPC	1789	M5
A2/10/RPPC	1766	M5
A2/20/RPPC	1254	M3
A2/30/RPPC	1730	M5
A2/5/RPP	1741	M5
A2/10/RPP	1749	M5
A2/20/RPP	1733	M5
A2/30/RPP	1720	M5

Table 26 - Classification of mortars according to NBR 13281:2005.

5.3.2 Modulus of elasticity

Table 27 shows the results of the modulus of elasticity tests. Table 28 shows the variation in modulus of elasticity between the mortars with residues and the reference mortars.

Mortar	Modulus of elasticity (GPa)	Standard deviation	Coefficient of variation (%)
A1/0/REF	7000	0,23	0,00
A1/5/RPPC	6500	0,20	0,00
A1/10/RPPC	6500	0,35	0,01
A1/20/RPPC	3900	0,00	0,00
A1/30/RPPC	5800	0,20	0,00
A1/5/RPP	6900	0,30	0,00
A1/10/RPP	6100	0,31	0,01
A1/20/RPP	6000	0,76	0,01

A1/30/RPP	5200	0,35	0,01
A2/0/REF	10600	0,26	0,00
A2/5/RPPC	9700	0,21	0,00
A2/10/RPPC	8400	0,23	0,00
A2/20/RPPC	8100	0,15	0,00
A2/30/RPPC	5900	0,25	0,00
A2/5/RPP	8700	0,25	0,00
A2/10/RPP	9000	0,15	0,00
A2/20/RPP	8000	0,25	0,00
A2/30/RPP	6500	0,61	0,01

Table 27 - Average results of the modulus of elasticity of the mortars.

Mortar	fcal/cim	fagr/agl	fres/cim	t (μs)	Variation between the modulus of elasticity of mortars with residues and reference mortars (%)
A1/5/RPPC	0,66	5,48	0,06	79,5	-7,14
A1/10/RPPC	0,70	5,66	0,12	79,0	-7,14
A1/20/RPPC	0,79	6,06	0,27	67,1	-44,29*
A1/30/RPPC	0,90	6,51	0,46	83,0	-17,14
A2/5/RPPC	0,33	5,14	0,06	65,3	-8,49
A2/10/RPPC	0,35	5,35	0,12	69,6	-20,75
A2/20/RPPC	0,39	5,83	0,27	71,3	-23,58
A2/30/RPPC	0,45	6,40	0,46	82,9	-44,34
A1/5/RPP	0,66	5,48	0,05	76,9	-1,43
A1/10/RPP	0,70	5,66	0,11	81,9	-12,86
A1/20/RPP	0,79	6,06	0,24	82,4	-14,29
A1/30/RPP	0,90	6,51	0,41	88,7	-25,71
A2/5/RPP	0,33	5,14	0,05	68,10	-17,92
A2/10/RPP	0,35	5,35	0,11	67,3	-15,09
A2/20/RPP	0,39	5,83	0,24	70,6	-24,53
A2/30/RPP	0,45	6,40	0,41	78,9	-38,68

*The value should be disregarded. In this case, there must have been a fault when the test was carried out.

Table 28 - Variation in modulus of elasticity between mortars with residues and reference mortars.

As already discussed in Chapter 3, the modulus of elasticity is of fundamental importance for masonry mortars. The importance of this property in mortars is directly related to their capacity for deformation and its link to pathological phenomena, such as detachment of the cladding.

The results shown in Table 28 show that as the lime/cement, residue/cement and aggregate/binder ratios increase, the modulus of elasticity decreases for mortars produced with RPPC and RPP when compared to conventional mortars (produced without residues).

With the decrease in cement content caused by the increase in the lime/cement, residue/cement and aggregate/binder ratios, the compressive strength, flexural tensile strength and tensile bond strength probably decrease and, consequently, the modulus of elasticity decreases.

The propagation time of the ultrasonic wave is directly proportional to the lime/cement, waste/cement and aggregate/binder ratios, according to the results shown in Table 28. As the compactness of the binder and aggregate increases, most likely due to the increase in the fines content (lime and powdery material), the propagation speed of the ultrasonic wave increases and, consequently, the propagation time of the wave decreases, thus characterizing the increase in the modulus of elasticity.

The mortars produced with the A1 mix have a higher modulus of elasticity than those made with the A2 mix, both produced with RPPC. This can be explained by the fact that the specific mass of the mixture is higher for mix A1. Possibly it has a higher modulus of elasticity. The same occurred for traits A1 and A2, produced with RPP.

The correlation (Figure 34) between the traits showed that the decrease in the modulus of elasticity is directly proportional to the increase in the percentage of residue present in the mixture, regardless of the trait or residue used.

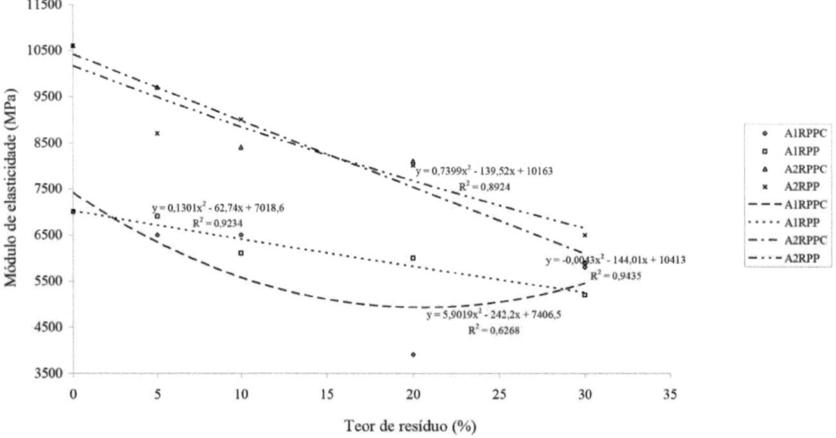

Figure 34 - Correlation between residue content and the elastic modulus of mortars.

5.3.3 Tensile strength in bending

The average results of the flexural tensile strength test are shown in Table 29. The results of the variation in flexural tensile strength between the mortars with residues and the reference mortars are shown in Table 30.

Mortar	Flexural tensile strength (MPa)	Standard deviation	Coefficient of variation (%)
A1/0/REF	1,08	0,06	5,46
A1/5/RPPC	1,03	0,04	3,78
A1/10/RPPC	1,10	0,05	4,45
A1/20/RPPC	0,39	0,05	13,25
A1/30/RPPC	0,61	0,27	44,79
A1/5/RPP	1,02	0,12	11,30
A1/10/RPP	0,86	0,06	6,65
A1/20/RPP	0,71	0,03	4,65
A1/30/RPP	0,72	0,18	25,36
A2/0/REF	1,85	0,12	6,37
A2/5/RPPC	1,69	0,04	2,56
A2/10/RPPC	1,42	0,06	4,19
A2/20/RPPC	1,46	0,08	5,27
A2/30/RPPC	0,76	0,20	25,96
A2/5/RPP	1,75	0,15	8,77
A2/10/RPP	1,39	0,02	1,67
A2/20/RPP	1,38	0,13	9,16
A2/30/RPP	0,75	0,32	42,14

Table 29 - Average results of the mortars' flexural tensile strength.

Mortar	$f_{lime/cim}$	$f_{agr/agl}$	$f_{res/cim}$	Variation between the flexural tensile strength of mortars with residues and reference mortars (%)
A1/5/RPPC	0,66	5,48	0,06	-4,63
A1/10/RPPC	0,70	5,66	0,12	+1,85
A1/20/RPPC	0,79	6,06	0,27	-63,89*
A1/30/RPPC	0,90	6,51	0,46	-43,52
A1/5/RPP	0,66	5,48	0,05	-5,56
A1/10/RPP	0,70	5,66	0,11	-20,37
A1/20/RPP	0,79	6,06	0,24	-34,26
A1/30/RPP	0,90	6,51	0,41	-33,33
A2/5/RPPC	0,33	5,14	0,06	-8,65
A2/10/RPPC	0,35	5,35	0,12	-23,24
A2/20/RPPC	0,39	5,83	0,27	-21,08
A2/30/RPPC	0,45	6,40	0,46	-58,92
A2/5/RPP	0,33	5,14	0,05	-5,41
A2/10/RPP	0,35	5,35	0,11	-24,56

| A2/20/RPP | 0,39 | 5,83 | 0,24 | -25,41 |
| A2/30/RPP | 0,45 | 6,40 | 0,41 | -59,46 |

*The value should be disregarded. In this case, there must have been a fault when the test was carried out.

Table 30 - Variation in flexural tensile strength between mortars with residues and reference mortars.

When analyzing the results shown in Table 30, it can be seen that as the lime/cement, residue/cement and aggregate/binder ratios increase, there is a significant decrease in flexural tensile strength. In addition, the mortars studied had an apparent mass density in the hardened state in the higher ranges of NBR 13281:2005, contrary to its recommendations. As such, they may be contributing negatively to the mechanical performance of the mortars formulated, since, according to Silva et al (2007), the bulk density in the hardened state is an important property that can explain the coating's performance in terms of mechanical properties, influenced mainly by the packing of the particles of the fine aggregates and also by the packing of the aggregate/binder combination.

In this study, cement was partially replaced by a percentage of waste, and possibly because of this, the flexural tensile strength performance was reduced in all the mortars produced with waste. It was observed that the mortars produced with the A1 mix performed even more poorly than those produced with the A2 mix. This may have been due to the fact that the A1 mix has a 50% higher lime/cement ratio than the A2 mix. Therefore, the partial replacement of cement with waste may have compromised the performance of the mortars formulated with this mix.

It should be noted that the mortars used for cladding are associated with flexural tensile strength. Low flexural tensile strength does not allow the mortar to withstand the stresses, which can lead to pathologies such as detachment of the cladding.

An analysis of the correlations (Figure 35) shows that an increase in the residue content when producing mortars tends to reduce their mechanical strength.

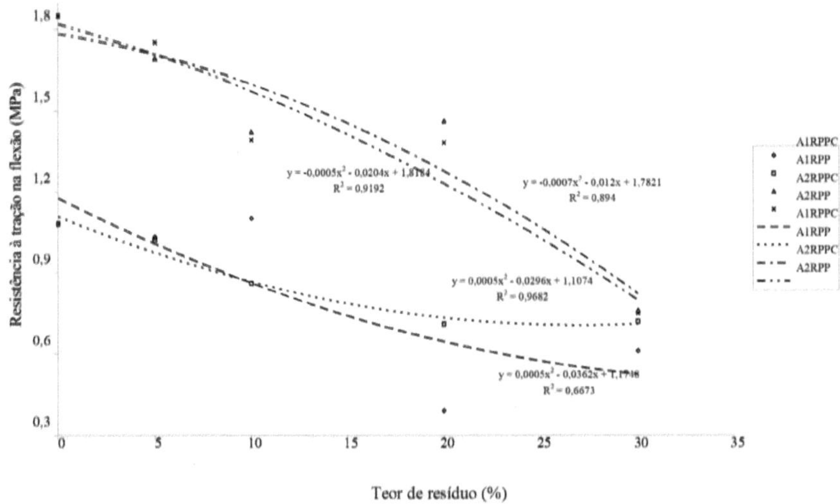

Figure 35 - Correlation between residue content and the flexural tensile strength of mortars.

The classification of the mortars studied, according to the criteria established by NBR 13281 (ABNT, 2005), Annex C, is shown in Table 31. The standard emphasizes that mortars with good performance and durability are classified in the higher ranges of the table. The results clearly show that in relation to flexural tensile strength, A1 mortars are classified in the lowest range of the table.

It should be noted that there was no difference in classification between the mortars produced with the same mix and type of residue when the percentage of residue present in the mixture was increased.

Mortar	Tensile strength in bending (MPa)	Classification - NBR 13281:2005
A1/0/REF	1,08	R1
A1/5/RPPC	1,03	R1
A1/10/RPPC	1,10	R1
A1/20/RPPC	0,39	R1
A1/30/RPPC	0,61	R1
A1/5/RPP	1,02	R1
A1/10/RPP	0,86	R1
A1/20/RPP	0,71	R1
A1/30/RPP	0,72	R1
A2/0/REF	1,85	R3
A2/5/RPPC	1,69	R3

A2/10/RPPC	1,42	R2
A2/20/RPPC	1,46	R2
A2/30/RPPC	0,76	R1
A2/5/RPP	1,75	R3
A2/10/RPP	1,39	R2
A2/20/RPP	1,38	R2
A2/30/RPP	0,75	R1

Table 31 - Classification of mortars according to NBR 13281:2005.

5.3.4 Water absorption by capillarity

Tables 32 and 33 show the results of the mortars studied in relation to water absorption by capillarity.

Mortar	Absorption of water by capillarity - t = 90 minutes $(g/cm)^2$	Standard deviation	Coefficient of variation (%)
A1/0/REF	1,46	0,06	4,28
A1/5/RPPC	1,70	0,01	0,59
A1/10/RPPC	1,62	0,05	3,05
A1/20/RPPC	1,85	0,05	2,78
A1/30/RPPC	2,08	0,04	2,10
A1/5/RPP	1,61	0,03	2,15
A1/10/RPP	1,52	0,03	2,28
A1/20/RPP	1,63	0,05	3,25
A1/30/RPP	1,20	0,04	3,33
A2/0/REF	0,66	0,23	34,85
A2/5/RPPC	1,17	0,02	1,77
A2/10/RPPC	0,99	0,19	19,16
A2/20/RPPC	1,38	0,18	13,30
A2/30/RPPC	1,58	0,13	8,52
A2/5/RPP	0,89	0,14	15,28
A2/10/RPP	0,93	0,04	3,79
A2/20/RPP	1,13	0,13	11,71
A2/30/RPP	1,29	0,02	1,18

Table 32 - Average results of water absorption by capillarity (t = 90 minutes) of the mortars.

Mortar	Variation between water absorption by capillarity at 90 minutes of mortars with residues and reference mortars (%)
A1/5/RPPC	+16,44
A1/10/RPPC	+10,96

A1/20/RPPC	+26,71
A1/30/RPPC	+42,47
A1/5/RPP	+10,27
A1/10/RPP	+4,11
A1/20/RPP	+11,64
A1/30/RPP	+17,81*
A2/5/RPPC	+77,27
A2/10/RPPC	+50,00
A2/20/RPPC	+109,09
A2/30/RPPC	+139,39
A2/5/RPP	+34,85
A2/10/RPP	+40,91
A2/20/RPP	+71,21
A2/30/RPP	+95,45

*The value should be disregarded. In this case, there must have been a fault when the test was carried out.

Table 33- Variation in water absorption by capillarity at 90 minutes between mortars with residues and reference mortars.

Table 33 shows that there was an increase in water absorption in all the mortars formulated with residues. The mortars with RPPC in their mix showed even greater absorption when compared to those produced with RPP, regardless of the type of mix used. This may be due to the higher content of fines in the RPPC, which allows the formation of pores with smaller diameters and possibly greater connectivity between them. It should be noted that the height of water penetration is inversely proportional to the diameter of the capillaries, i.e. the smaller the diameter of the capillaries, the greater the height the water will reach in the test specimen.

The correlations show (Figure 36) that there is a tendency for mortars to increase water absorption by capillarity when the residue content in their mixtures increases.

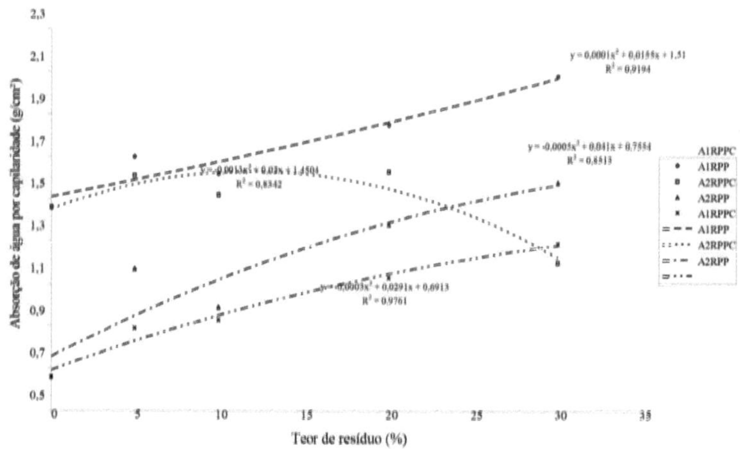

Figure 36 - Correlation between residue content and water absorption by capillarity (t = 90 minutes) of mortars.

5.3.5 Microstructural analysis

Figure 37 illustrates the reference A1 trace where we can see the formation of ettringite crystals (1), capillary voids (2) and hydrated calcium silicates (3). Figure 38 illustrates the spectroscopic analysis carried out on the central area of Figure 37.

Figure 37 - Micrograph of the reference A1 trace.

Figure 38 shows the main components of the mixture are Ca, Si, Al and O. The predominance of Ca is due to the presence of lime in the mortar. The reaction of Ca with Si is responsible for the formation of hydrated calcium silicate. This phase makes up 50 to 60% of the volume of solids in a fully hydrated Portland cement matrix. It is also the most important phase in determining the mortar's properties. The presence of Al contributed to the formation of ettringite. It should be noted that ettringite contains small amounts of iron oxide, which can replace aluminum oxide in the crystal structure.

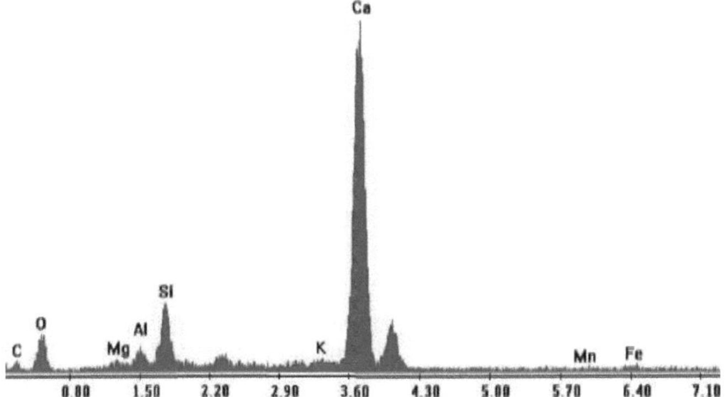

Figure 38 - Spectroscopic analysis carried out on reference trace A1.

Figure 39 illustrates trace A1 produced with 30% RPPC. The presence of capillary voids (1), the formation of ettringite needles (2), calcium hydroxide (3) and hydrated calcium silicates (4) can be seen. The spectroscopic analysis of this trace is shown in Figure 40. It was carried out on the central area of Figure 39.

Figure 39 - Micrograph of mix A1 produced with 30% RPPC.

Figure 40 shows the predominance of Ca, due to the presence of lime in the mixture. Ca and Si are responsible for the formation of hydrated calcium silicate. Al is contributing to the formation of ettringite. Ca and O are reacting and contributing to the formation of calcium hydroxide, this phase makes up 20 to 25% of the volume of solids in a Portland cement matrix.

87

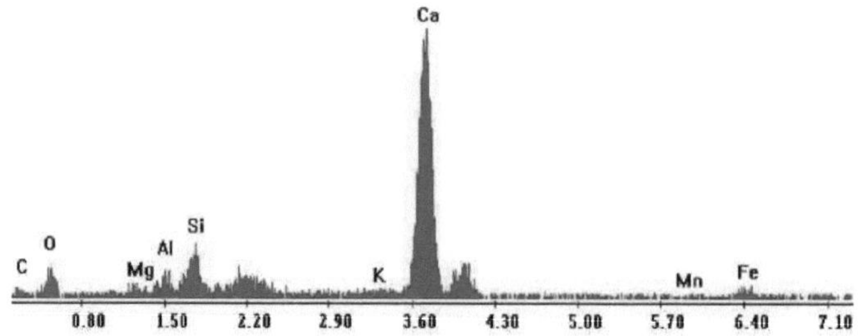

Figure 40 - Spectroscopic analysis of the Al trace produced with 30% RPPC.

Figure 41 illustrates trace A1 produced with 30% RPP. It clearly illustrates the presence of capillary pores (1) and hydrated calcium silicates (2). Figure 42 shows the spectroscopic analysis carried out on the central area of Figure 41.

Figure 41 - Micrograph of mix A1 produced with 30% RPP.

In Figure 42, the spectroscopic analysis shows the presence of Ca and Si, which, when they react, confirm the formation of the hydrated calcium silicates found in the microstructural analysis (Figure 41).

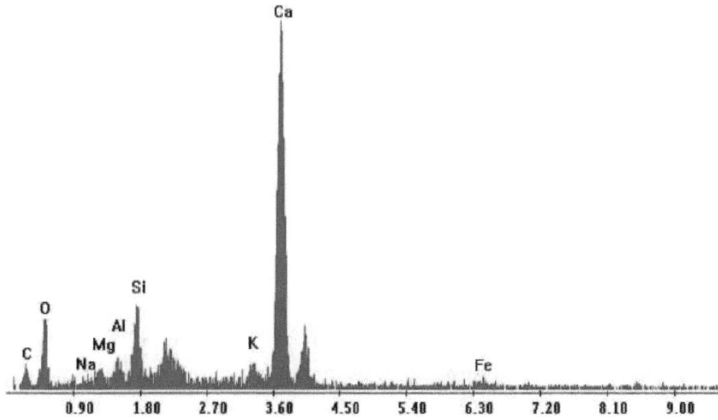

Figure 42 - Spectroscopic analysis of the Al trace produced with 30% RPP.

Figure 43 shows unhydrated cement grains (1), capillary voids (2), hydrated calcium silicates (3), calcium hydroxide (4) and the formation of ettringite needles (5). Figure 44 illustrates the spectroscopic analysis of the A2 reference trace carried out in the central area of Figure 43.

Figure 43 - Micrograph of reference trace A2.

Figure 44 shows the predominant presence of Ca, Si, Al and O in the mixture. Ca and Si are responsible for the formation of calcium silicate. Ca and O contribute to the formation of calcium hydroxide and Al to the formation of ettringite crystals.

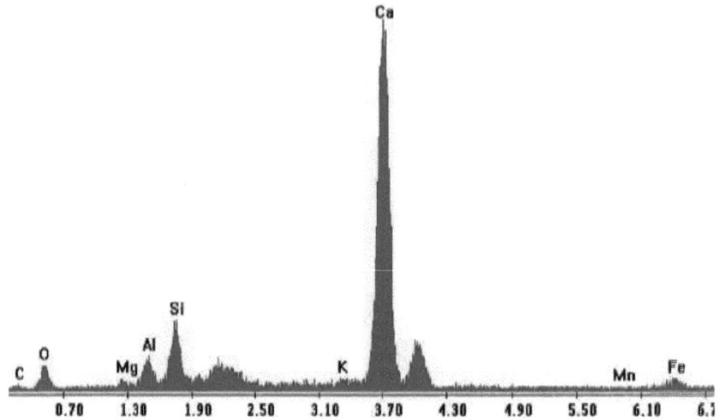

Figure 44 - Spectroscopic analysis of reference trace A2.

Figure 45 illustrates the micrograph taken of the A2 mix with 30% RPPC. The analysis shows the formation of hydrated calcium silicates (1), ettringite crystals (2), calcium hydroxide (3) and capillary voids (4) and the spectroscopic analysis carried out on the central area of this figure is shown in Figure 46.

Figure 45 - Microscopy of trace A2 produced with 30% RPPC.

In Figure 46, the spectroscopic analysis shows the predominant presence of Ca, Si, O and Al. Al is contributing to the formation of ettringite crystals. When they react, Ca and Si contribute to the formation of hydrated calcium silicates. Ca and O contribute to the formation of calcium hydroxide.

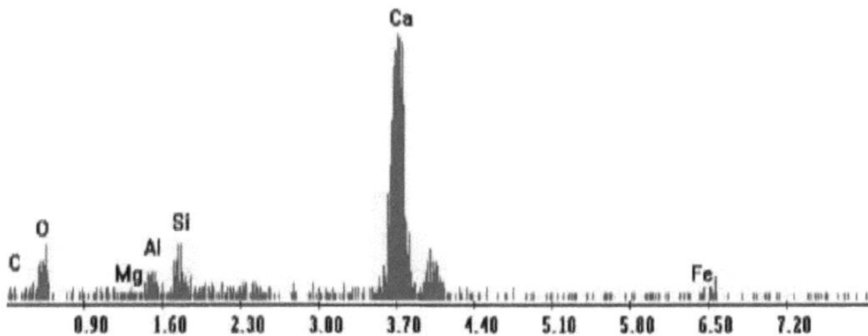

Figure 46 - Spectroscopic analysis of trace A2 produced with 30% RPPC.

Figure 47 illustrates the micrograph of the A2 mix produced with 30% RPP. It shows the formation of hydrated calcium silicates (1), capillary voids (2), ettringite needles (3) and calcium hydroxide (4). The spectroscopy of this trace was carried out in the central area of Figure 47 and is illustrated in Figure 48.

Figure 47 - Micrograph of mix A2 produced with 30% RPP.

The spectroscopic analysis (Figure 48) shows that Ca, Si, O and Al are the predominant elements. The formation of hydrated calcium silicates is due to the presence of Ca and Si. Al is contributing to the formation of ettringite and the reaction of Ca and O to the formation of calcium hydroxide.

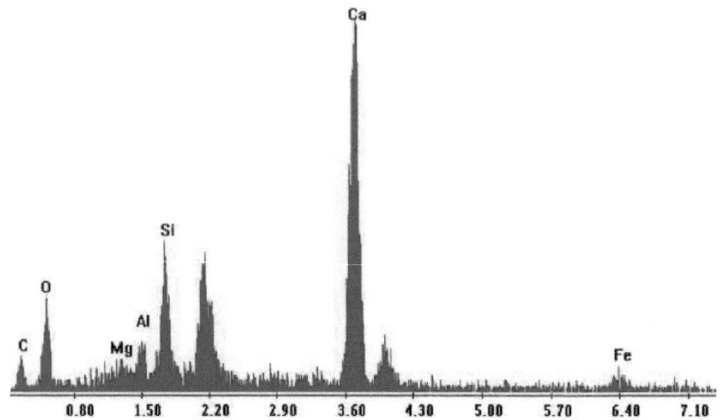

Figure 48 - Spectroscopic analysis of trace A2 produced with 30% RPP.

The microstructural analysis of the mortars showed that partially replacing cement with a percentage of waste, regardless of the type, influences the microstructure of the mortars. Mortars produced with RPPC had a denser microstructure than those produced with RPP, regardless of the mix used. This is due to the fact that RPPC has a higher specific mass and consequently a higher content of powdery material than RPP. The greatest crystal formation occurred in the mortars produced with the A2 mix, possibly because it has a lower lime/cement ratio than the A1 mix. No RPPC or RPP particles were identified in the analysis. They may have reacted with the lime, suggesting that the waste studied has pozzolanic activity. Scanning electron microscopy analysis showed greater crystal formation in the A2 traces, regardless of the type of waste used.

The spectroscopic analysis carried out at some points identified the different constituent elements of the mortars. It can be seen that there were no differences between the constituents of the mortars studied.

5.3.6 X-ray diffractometry and X-ray fluorescence analysis

Figure 49 shows the X-ray diffractometry analysis for the reference A1 mix. The analysis shows the predominance of quartz, due to the presence of sand in the mortar, as well as calcium hydroxide and calcite, due to the presence of lime in the mix.

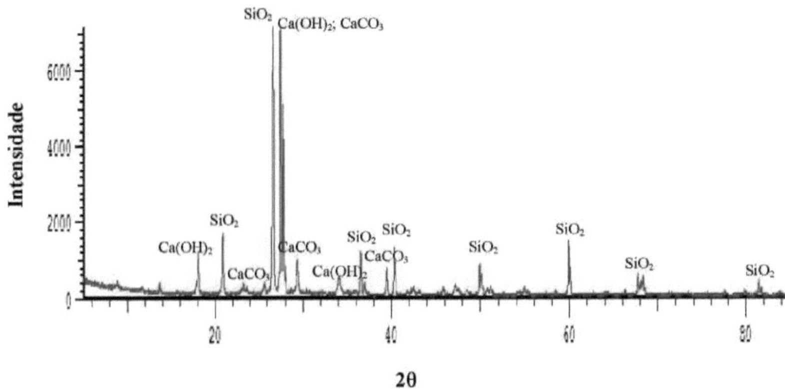

Figure 49 - XRD of reference trace A1.

Table 34 shows the results of the X-ray fluorescence analysis of the A1 reference trace. The analysis shows Ca and Si as the main components of the trace. Ca contributes to the formation of calcium hydroxide, calcium carbonate and calcite. Si comes from the presence of sand in the mortar.

Phase	Percentage
SiO2	43,311%
CaO	39,611%
Al2O3	8,944%
SO3	3,581%
Fe2O3	2,057%
K2O	1,856%
Others	0,640%

Table 34 - X-ray fluorescence of reference trace A1.

Figure 50 shows the main phases formed by trace A1 with 30% RPPC. X-ray diffractometry analysis shows the formation of calcium hydroxide, calcite and quartz.

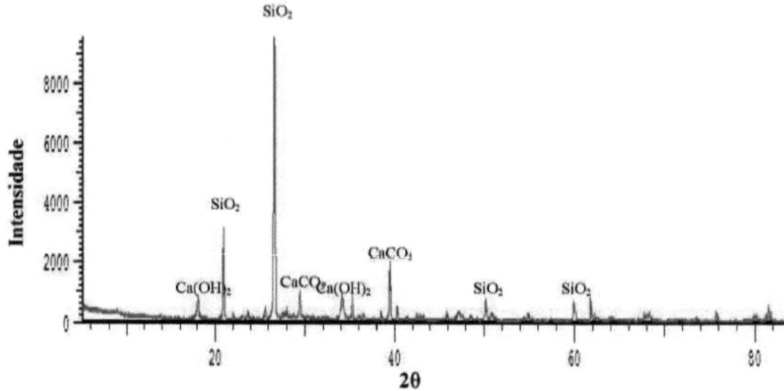

Figure 50 - XRD of trace A1 with 30% RPPC.

In Table 35, the X-ray fluorescence analysis shows that Ca contributes to the formation of the calcium hydroxide and calcite phases found in the XRD analysis (Figure 51). The presence of Si is due to the sand being one of the components of the mortar.

Phase	Percentage
SiO2	45,023%
CaO	38,978%
Al2O3	8,055%
SO3	3,329%
Fe2O3	2,014%
K2O	1,808%
Others	0,793%

Table 35 - X-ray fluorescence of trace A1 with 30% RPPC.

Figure 51 shows the X-ray diffractometry analysis of the A1 mix with 30% RPP. The analysis shows the presence of quartz, calcium hydroxide and calcite in the mortar.

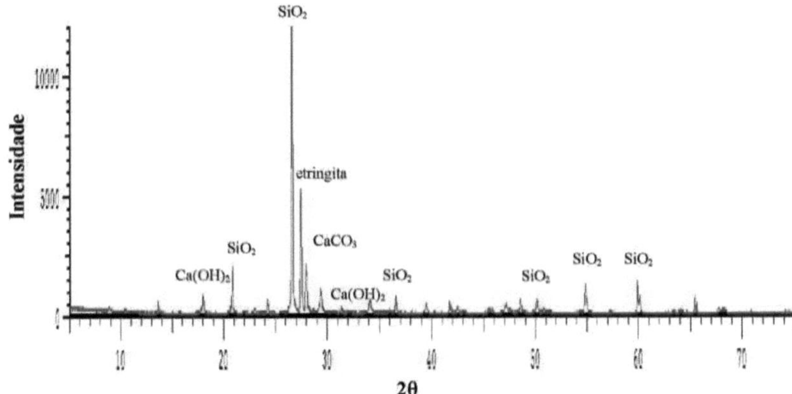

Figure 51 - XRD of trace A1 with 30% RPP.

Table 36 shows the results of the X-ray fluorescence analysis of the A1 mix with 30% RPP. The Si is due to the presence of sand in the mortar. Ca contributes to the formation of calcium hydroxide and calcite, and together with Al O_{23} and Fe O_{23} in the formation of ettringite crystals.

Phase	Percentage
SIO2	42,150%
CaO	41,017%
Al2O3	8,557%
SO3	3,365%
Fe2O3	2,127%
K2O	1,873%
Others	0,911%

Table 36 - X-ray fluorescence of A1 with 30% RPP.

X-ray diffractometry analysis of the A2 reference trace (Figure 52) shows quartz, ettringite, calcium hydroxide and calcite as the predominant phases.

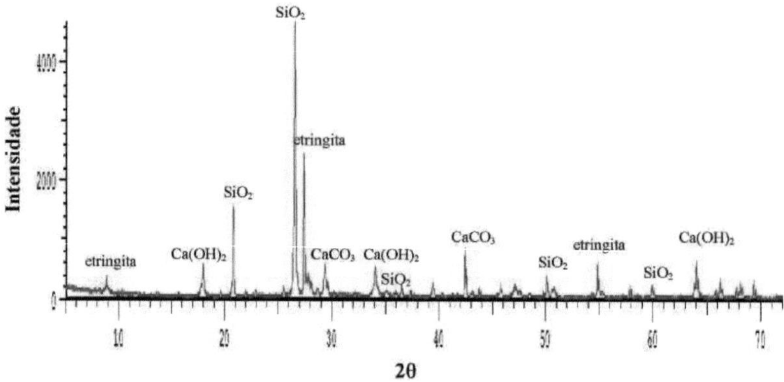

Figure 52 - XRD of trace A2 REF.

Table 37 shows the results of the X-ray fluorescence analysis of the A2 reference trace. The analysis shows Ca and Si as the predominant components. The Si is due to the presence of sand in the mix. Ca contributes to the formation of calcium hydroxide and calcite, and together with Al O_{23} and Fe O_{23} contributes to the formation of ettringite crystals.

Phase	Percentage
CaO	42,102%
SiO_2	41,426%
Al2O3	7,893%
SO3	3,960%
Fe2O3	2,315%
K2O	1,467%
Others	0,837%

Table 37 - X-ray fluorescence of trace A2 REF.

Figure 53 shows the X-ray diffractometry analysis of trace A2 with 30% RPPC. The analysis shows quartz, ettringite, calcite and calcium hydroxide.

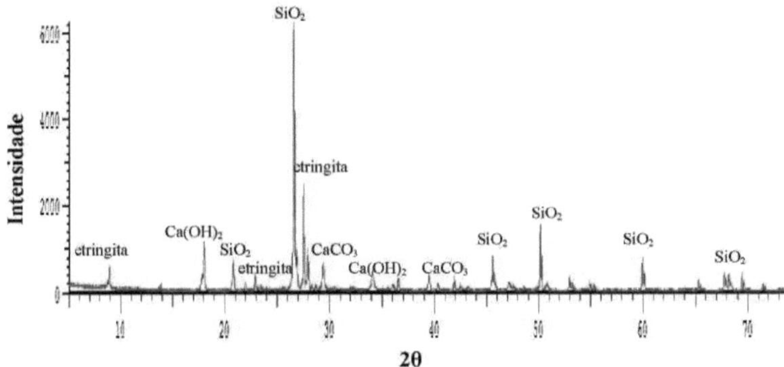

Figure 53 - XRD of trace A2 with 30% RPPC.

Table 38 shows the results of the X-ray fluorescence analysis of trace A2 with 30% RPPC. Si and Ca are the predominant components of the mixture. Ca contributes to the formation of calcium hydroxide and calcite, and together with $Al\ O_{23}$ and $Fe\ O_{23}$ contributes to the formation of ettringite crystals. The Si is due to the fact that sand is present in the mortar.

Phase	Percentage
SiO2	44,051%
CaO	38,934%
Al2O3	8,644%
SO3	3,591%
Fe2O3	2,179%
K2O	1,769%
Others	0,832%

Table 38 - X-ray fluorescence of trace A2 with 30% RPPC.

Figure 54 shows the results of the X-ray diffractometry analysis of trace A2 with 30% RPP. The analysis shows a predominance of quartz, ettringite, calcite and calcium hydroxide.

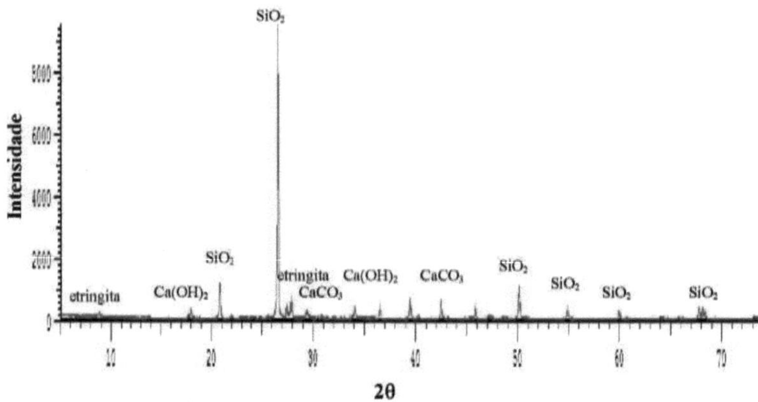

Figure 54 - XRD of trace A2 with 30% RPP.

Table 39 shows the results of the X-ray fluorescence analysis of the A2 30% RPP trace. The analysis shows that Si and Ca are the predominant components of the trace element. Ca contributes to the formation of calcite and calcium hydroxide, and together with Al O_{23} and Fe O_{23} contributes to the formation of ettringite crystals. Si comes from the presence of sand in the mix.

Phase	Percentage
SiO_2	49,833%
CaO	31,368%
Al2O3	9,904%
SO3	3,236%
K2O	1,901%
Fe2O3	1,821%
MgO	1,227%
Others	0,710%

Table 39 - X-ray fluorescence of trace A2 with 30% RPP.

X-ray diffractometry and X-ray fluorescence analyses show that the phases formed (quartz, calcium hydroxide, ettringite and calcite) when cement was partially replaced by residues in the formulated mortars are similar to the phases formed in the reference mortars. Therefore, the presence of residues in the mixture probably does not alter the phases of conventional mortars. These analyses confirm the results found in the micrographs and spectroscopic analyses.

CHAPTER 6

FINAL CONSIDERATIONS

6.1 CONCLUSIONS

This chapter contains the main conclusions reached in the Experimental Program of the research covered in this book, as well as suggestions for future work. It should be emphasized that the results obtained should be confirmed through further research, which will present results that can complement those obtained in this study.

As for the properties of the mortars in the fresh state, it was observed that:

The presence of the residue in the mix when the mortars were produced reduced the consistency of the mortars, and consequently increased their cohesion. It should also be noted that the mortars produced with RPPC had their consistency reduced by a maximum of 9%. When comparing the consistency indices of mortars with the same mix in relation to the type of residue used in their formulation, it can be seen that the mortars produced with RPPC had a lower consistency index than those produced with RPP. This loss in consistency and gain in cohesion could lead to gains in quality and productivity when the mortars are used. The correlations show that there is a tendency for mortars to decrease their consistency index when the percentage of residue present in the mixture increases.

As for water retention, the presence of the residue hardly altered this property, regardless of the type of residue and the mix used. However, by correlating the percentage of residue present in the mixture with the percentage of water retained, it can be seen that all the mortars produced with residue tend to increase water retention as the residue content in their composition increases. In terms of classification, the results show that the mortars studied comply with NBR 13281:2005, since they were classified in the higher ranges of the table, as determined by the standard, and that there is no relevant difference between the mortars formulated with residue and the reference mortars.

The results of the research show that the presence of residue in the mortar mix did not alter the density of the mortar, regardless of the type of residue and the mortar mix used. However, for the mortars produced with RPPC, correlating the percentage of residue present in the mixtures with the bulk density showed a tendency for the density to increase. With regard to classification, the results show that there is no variation between the mortars formulated with residue and the reference mortar and that the mortars are in the central range of the classification table established by the standard.

With regard to the content of incorporated air, the results show that this property does not have its performance altered by the presence of the residue in the mortar mixtures. However, there is a tendency for the incorporated air content of mortars produced with RPP to increase when compared to those produced with RPPC.

Regarding the properties of mortars in the hardened state, it was observed that:

The presence of residue in the mix, regardless of its type or content, did not change the mass density of the mortars formulated. However, the correlations found between the traits studied show that the mortars produced with residues tend to reduce their bulk density as the residue content in the mixture increases. As for classification, it was found that there is no difference in classification between the mortars formulated with waste and the reference mortar. However, the mortars are in the high range of the classification table. In view of this, their performance and durability could possibly be compromised when they are applied.

In relation to the modulus of elasticity of the mortars, the tests showed that the higher the residue/cement ratio, the lower the modulus of elasticity, since the mortars dosed with residue had a lower speed of propagation of the ultrasonic wave and consequently a longer wave propagation time, thus reducing the modulus of elasticity. The mortars produced with RPPC had the lowest modulus of elasticity. The correlations indicate that there is a tendency for the mortars formulated to reduce their modulus of elasticity when the residue content in the mix is increased, regardless of the mix or residue used.

The mortars produced with waste showed compromised performance in terms of flexural tensile strength. The mortars produced with the A1 mix showed a more unfavorable decrease than those produced with the A2 mix. In relation to the type of residue used in the mix, there was no significant difference between those produced with RPPC and RPP. The correlations indicate that the higher the residue content in the mix, the lower the mechanical performance of the mortars. In terms of classification, the results clearly show that in relation to flexural tensile strength, A1 mortars are classified in the lowest range of the table and A2 mortars in the intermediate ranges. It should be noted that there was no difference in classification between the mortars produced with the same mix and type of residue when the percentage of residue present in the mixture was increased.

The results showed that the property of water absorption by capillarity was also impaired when cement was replaced by residue in the mortar formulation. The mortars produced with RPPC had the highest water absorption, regardless of the mix used. The correlations indicate that water absorption by capillarity is directly proportional to the residue content.

As for the microstructural analysis, it was not possible to identify particles of RPPC and RPP. They probably reacted with the lime, suggesting that the waste had pozzolanic activity. The micrographs showed that the mortars produced with RPPC have a denser microstructure than those produced with RPP. The mortars produced with the A2 trait showed greater crystal formation than those produced with A1. Spectroscopic analysis showed that there was no difference between the constituents of the mortars studied.

The XRD and XRF analyses show that the phases formed in the mortars formulated with residues are similar to those formed in the reference mortars. Therefore, it is possible that the residues do not influence the formation of the mortar phases.

6.2 ORIGINAL CONTRIBUTION TO KNOWLEDGE

As an original contribution to knowledge it can be said that, possibly:

•   The mortars formulated and studied in this research can be applied as laying and coating mortars, respectively, when building masonry.

•   Waste can replace cement by up to 30% when formulating mortars.

•   *There is* no need to beneficiate the waste in order to use it in mortars.

•   The presence of residues in the mixtures reduced the modulus of elasticity of the mortars by more than 40%.

6.3 SUGGESTIONS FOR FUTURE WORK

Considering the scope of the subject and the fact that it is impossible to exhaust all the possibilities for studying the behavior of a material in a single study, especially when it comes to a new material, there are several studies that can be carried out in order to contribute to increasing knowledge on the subject. With this in mind, we suggest

•   Evaluate the effects of shrinkage of mortar with residues in the fresh and hardened state;

•   Evaluate the tensile strength of mortars with residues;

•   To evaluate the influence of additives in place of lime on the properties of mortars with residues;

•   Research into the correlation between modulus of elasticity and flexural tensile strength, compressive strength and tensile bond strength in the control of cracking in mortar coatings with residues;

•   Carry out a porosity study of mortars produced with waste for low water/cement ratios;

•   To study the effect of different types of curing on the properties and durability of mortars with residues;

- Carry out a more in-depth study on the characterization of the RPPC and RPP;

- Carry out a study on the use of RPPC or RPP together with recycled fine aggregate from construction and demolition waste in the production of mortars;

- Evaluate the absorption rate of mortars with residue;

- To analyze the microstructure of the mortars formulated with residues at ages of 3, 7 and 14 days, in order to verify the evolution of the reactions of the residues.

REFERENCES

AÏTCIN, P. C. **High performance concrete**. Sao Paulo: PINI, 2000.

AMORIM, L. V.; PEREIRA, A. S. G.; NEVES, G. de A.; FERREIRA, H. C. **Recycling construction waste for use in low-cost mortars. Revista Brasileira de engenharia agricola ambiental**. Vol. 3. n. 2. 1999. p. 222-228. Available at:

http://www.agriambi.com.br/revista/v3n2/222.pdf. Accessed on: June 27, 2008.

ANGELIM, R. R.; ANGELIM, S. C. M.; CARASEK, H. **Influence of the addition of calcareous, siliceous and clayey fines on the properties of mortars and coatings**. In: V Brazilian Symposium on Mortar Technology. Proceedings...CD-ROM. Sao Paulo: ANTAC/POLI-USP, 2003. p. 401-416.

ÂNGULO, S. C. **Characterization of recycled construction and demolition waste aggregates and the influence of their characteristics on concrete behavior**. Sao Paulo, 2005. 167 p. Thesis (Doctorate). Department of Civil Construction Engineering. Polytechnic School of the University of Sao Paulo. Available at: http://www.reciclagem.pcc.usp.br/ftp/caracterizacaoSA1.pdf. Accessed on: June 27, 2008.

BRAZILIAN ASSOCIATION OF TECHNICAL STANDARDS. NBR 7211 - **Aggregate for concrete**. Rio de Janeiro, 1983.

BRAZILIAN ASSOCIATION OF TECHNICAL STANDARDS. NBR 7215 - **Portland cement - determination of compressive strength**. Rio de Janeiro, 1996.

BRAZILIAN ASSOCIATION OF TECHNICAL STANDARDS. NBR 7217 - **Aggregates - determination of granulometric composition**. Rio de Janeiro, 1987.

BRAZILIAN ASSOCIATION OF TECHNICAL STANDARDS. NBR 7251 - **Loose aggregate - determination of unit mass**. Rio de Janeiro, 1982.

BRAZILIAN ASSOCIATION OF TECHNICAL STANDARDS. NBR 9776 - **Aggregates: determination of the specific mass of fine aggregates using Chapman's method**. Rio de Janeiro, 1987.

BRAZILIAN ASSOCIATION OF TECHNICAL STANDARDS. NBR 9778 - **Mortar and hardened concretes - determination of water absorption, voids index and specific mass**. Rio de Janeiro, 2006.

BRAZILIAN ASSOCIATION OF TECHNICAL STANDARDS. NBR 13276 - **Mortar for laying and coating walls and ceilings - preparation of the mixture and determination of the consistency index**. Rio de Janeiro, 2005.

BRAZILIAN ASSOCIATION OF TECHNICAL STANDARDS. NBR 13277 - **Mortar for wall laying and wall and ceiling cladding - determination of water retention**. Rio de Janeiro, 1995.

BRAZILIAN ASSOCIATION OF TECHNICAL STANDARDS. NBR 13278 - **Mortar for laying and coating walls and ceilings - determination of bulk density and incorporated air content**. Rio de Janeiro, 2005.

BRAZILIAN ASSOCIATION OF TECHNICAL STANDARDS. NBR 13279 - **Mortar for laying and coating walls and ceilings - determination of flexural tensile strength and compressive strength**. Rio de Janeiro, 2005.

BRAZILIAN ASSOCIATION OF TECHNICAL STANDARDS. NBR 13280 - **Mortar for laying and coating walls and ceilings - determination of mass density in the hardened state**. Rio de Janeiro, 2005.

BRAZILIAN ASSOCIATION OF TECHNICAL STANDARDS. NBR 13281 - **Mortar for laying and coating walls and ceilings - Requirements**. Rio de Janeiro, 2005.

BRAZILIAN ASSOCIATION OF TECHNICAL STANDARDS. NBR 13530 - **Inorganic mortar wall and ceiling coverings**. Rio de Janeiro, 1995.

BRAZILIAN ASSOCIATION OF TECHNICAL STANDARDS. NBR 15259 - **Mortar for laying and coating walls and ceilings - determination of water absorption by capillarity and the capillarity coefficient**. Rio de Janeiro, 2005.

BRAZILIAN ASSOCIATION OF TECHNICAL STANDARDS. CB-18/PROJETO 18:400.04-008 - **Mortar for laying and coating walls and ceilings - determination of the dynamic modulus of elasticity using ultrasonic wave propagation**. Rio de Janeiro, 2008.

BAIA, L. L.; SABBATINI, F. H. **Design and execution of mortar coatings**. Coleção primeiros passos da qualidade no site de obras. 1. ed. Sao Paulo: O Nome da Rosa, 2000. 82 p.

BARBOSA, K. C. **Experimental evaluation of the shrinkage phenomenon in concrete block masonry**. Sao Carlos, 2005. 233 p. Master's degree dissertation. Postgraduate Program in Civil Construction. Federal University of Sao Carlos. Available at: http://www.bdtd.ufscar.br/tde_arquivos/7/TDE-2005-04-1/T06:34:38Z- 592/Publico/DissKCB.pdf. Accessed on: July 5, 2008.

BAUER, E.; SOUSA, J. G. G.; GUIMARÂES, E. A. **Study of mortar consistency using the cone static penetration method**. In: VI Brazilian Symposium on Mortar Technology. Proceedings... CD ROM. Florianópolis: ANTAC, 2005. p. 95-105.

BERNARDINI, A. M.; FELISBERTO, D. S.; DAROS, M. T.; RIELLA, H.G. **Reuse of polishing and glazing residues to obtain cellular ceramics**. Ceràmica Industrial. Magazine. Vol. 11. n. 5. Sep/Dec, 2006. Sao Paulo/SP. Available at: http://www.ceramicaindustrial.org.br/pdf/v11n05/v11n5a05.pdf. Accessed on: July 7, 2008.

BORTOLUZZO, C. W.; LIBORIO, J. B. L. **Determination of stresses due to impeded shrinkage in coating mortars**. In: III Brazilian Symposium on Mortar Technology. **Proceedings...** CD ROM. Vitória: UEFS/PPGEC, 1999. Vol. 1. p. 125-136.

BRITISH STANDARDS INSTITUTION. BS 5628**: Part 3: materials and components, design and workmanship**. London, 1985.

BRIZOLA, R. M. **Microstructure of concrete coverings with high contents of slag and fly ash activated by Portland cement and hydrated lime**. Santa Maria, 2007. 178 p. Thesis (Doctorate). Postgraduate Program in Civil Engineering. Federal University of Santa Maria. Available at: http://coralx.ufsm.br/gepecon/Brizola.pdf. Accessed on: June 27, 2008.

BUDINSKI, K. G. **Engineering materials: porperties and selection**. 5. ed. New Jersey: Prentice Hall International, 1996. 653 p.

CALHAU, E. L. **Coating mortars with air-entraining additives - properties and recommendations**. Vitória, 2000. 144 p. Master's degree dissertation. Federal University of Espirito Santo.

CALLISTER, JR, W. D. **Fundamentals of materials science and engineering**. Rio de Janeiro: LCT, 2002. 589 p.

CALMON, J. L.; TRISTÂO, F. A.; LORDÊLLO, F. S. S.; DA SILVA, S. A. C.; MATTOS, F. V. **Utilization of granite cutting waste for the production of laying mortars**. In: II Brazilian Symposium on Mortar Technology. Proceedings... CD ROM. Salvador, 1997. p. 64-75.

CAPELLO, G. **Ecological cement - product has reduced CO_2 level in its formula, causing less impact on the environment**. Available at: http://planetasustentavel.abril.uol.com.br/inc/pop_print.html. Accessed on: June 20, 2008.

CARASEK, H. **Factors that influence the bond strength of mortars**. In: II Brazilian Symposium on Mortar Technology. Proceedings... CD ROM. Salvador: CEPED/EPUFBA/UCSAL/UEFS, 1997. p. 131-146.

CASCUDO, O.; CARASEK, H.; CARVALHO, A. **Control of industrialized mortars on site using the cone penetration method**. In: VI Brazilian Symposium on Mortar Technology. Proceedings... CD ROM. Florianópolis, May 2005. p. 8394.

CARVALHO Jr., A. N.; BRANDÂO, P. R. G.; FREITAS, J. M. C. **Relationship between the** adhesion strength of mortar coatings and the penetration profile of binder paste in the pores of ceramic blocks. In: VI Brazilian Symposium on Mortar Technology. Proceedings... CD ROM. Florianópolis: ANTAC, 2005. p. 518-529.

CAVANI, G. R.; ANTUNES, R. P. N.; JOHN, V. M. **Influence of incorporated air content on the workability of mixed mortars**. In: II Brazilian Symposium on Mortar Technology. Proceedings... CD ROM. Salvador: ANTAC, 1997. p. 110-119.

CEMENT.ORG. **Cement market in March 2008.** Available at: http://www.cimento.org/principal.htm. Accessed on: June 20, 2008.

CINCOTTO, M. A.; SILVA, M. A. C.; CASCUDO, H. C. **Coating mortars: characteristics, properties and test methods**. Boletim 68 IPT. Sao Paulo, 1995. 118 p.

COIMBRA, M. A.; LIBARDI, W.; MORELLI, M. R. **Utilization of zinc-coal pile tailings in Portland cement mortars and concretes**. Revista Cerâmica. Vol. 50. n. 316. p. 300-307. 2004. Available at: http://www.scielo.br/pdf/ce/v50n316/a0450316.pdf. Accessed on: June 29, 2008.

CRUZ, A. A. de F. **Study on the use of biomass burning residue in mortars**. Joinville, 2002. 99 p. Postgraduate Program in Materials Science and Engineering Santa Catarina State University. Available at: http://www.tede.udesc.br/tde_busca/arquivo.php?codArquivo.340. Accessed on: June 27, 2008.

D'AGOSTINO, L. Z.; SOARES, L. **Preparation of mortars using quarry fines**. In: IV Seminar on Sustainable Development and Recycling in Civil Construction - Recycled Materials and their Applications. Proceedings... Sao Paulo: IBRACON/CT-206, 2001. p. 375-386.

D'AGOSTINO, L. Z.; SOARES, L. **The use of granitic- gneissic rock quarry fines as a substitute for natural sands in the production of mortar**. Geosciences magazine. Vol. 22. n. 1. p. 65-73. Sao Paulo, 2003. State University of Sao Paulo. Available at: http://petro.rc.unesp.br/revistageociencias/22_1/6.PDF. Accessed on: June 29, 2008.

DEL CARLO, U. **Sustainable culture**. Téchne magazine. Ed. 133. Sao Paulo: PINI, 2008. p. 22-28.

DETRICHE, C. H.; RAMODA, S. **Influence of the volume composition of mortars on their resistance to crack propagation**. Materials and structures/Matériaux ET constructions. Vol. 22, n. 129. 1989. p. 206-211.

FARIAS FILHO, J. de; LUCENA, L. de F. L.; NEVES, G. de A.; FERREIRA, H. C.; LEAL, A. F. **Evaluation of the pozzolanic activity of civil construction waste**. 17th CBECIMat - Brazilian Congress of Materials Science and Engineering. Foz do Iguaçu, 2006. Available at:

http://www.metallum.com.br/17cbecimat/resumos/17cbecimat-112-007.pdf. Accessed on: June 27, 2008.

FIORITO, A. J. S. I. **Manual de argamassas e revestimento**: estudos e procedimentos de execuçao. Sao Paulo: PINI, 2003. 223 p.

GARCIA, T. L.; NAVES, V. C.; ALVES, J. D.; BARROS, B. de C.; RODRIGUES, C. S. G.; PEREIRA, E. A. A.; ATAiDES, R. L. de A. **Replacing part of the cement with inert quarry fines in concretes with a low w/c factor.** IV Scientific Initiation Seminar. Goiânia: 2006. Agricultural Engineering Course. University Unit of Exact and Technological Sciences, State University of Goiás. Available at:

http://www.prp.ueg.br/06v1/ctd/pesq/inic_cien/eventos/sic2006/arquivos/engenharias/substitu içao_cimento.pdf. Accessed on: June 28, 2008.

GOLASZEWSKI, J. **Interaction between superplasticizing cement and granulated blast furnace slag.** e-Mat - Journal of Science and Technology of Building Materials. Vol. 2. n. 1. p. 105-116. Nov. 2005. Available at: http://www.antac.org.br/emat/e-MAT-V2-N2/e-MAT-V2-N2-p105-116(Res).pdf. Accessed on: June 27, 2008.

GOMES, A. M.; NERO, J. M. G.; APPLETON, J. A. S. **New method for evaluating the workability and consistency of mortars.** In: I Brazilian Symposium on Mortar Technologies. Proceedings... CD ROM. Goiânia: ANTAC, 1995. p. 83-91.

GONDIM, F. L. **Study of the interface between concrete and repair material using microscopy.** Uberlândia, 2005. 143 p. Master's degree dissertation. Postgraduate Program in Civil Engineering. Federal University of Uberlândia. Available at: HTTP://www.bdtd.ufu.br//tdebusca/arquivo.php?codArquivo=227. Accessed on: July 29, 2008.

GUIMARÂES, J. E. P. **Lime - Fundamentals and Applications in Civil Engineering.** 2. ed. Sao Paulo: PINI, 2002. 341 p.

INTERNATIONAL UNION OF TESTING AND RESERCH LABORATORIES FOR MATERIALS AND STRUCTURES - RILEM. **MR 3: the complex workability - consistency - plasticity.** France: E & FN Spon. RILEM, 1982.

JOHN, V. M. **Reciclagem de residuos na construção civil: contribuçâo à metodologia de pesquisa e desenvolvimento.** Sao Paulo, 2000. 113 p. Thesis (Doctorate). Department of Civil Construction Engineering. Polytechnic School of the University of Sao Paulo.

JOHN, V. M. **Rethinking the role of hydrated lime in mortars.** In: V Brazilian Symposium on Mortar Technology. Proceedings ... CD ROM. Sao Paulo: ANTAC, 2003. p. 47-63.

KOLIAS, S. **Investigation of the possibility of estimating concrete strength by porosity measurements**. Materials and structures. Vol. 27. 1994. p. 265-272.

KOPSCHITZ, P.; FRANCINETE, Jr., P.; CINCOTTO, M. A.; JOHN, V. M. **Study of shrinkage and development of mechanical properties of mixed mortars for coating**. In: II Brazilian Symposium on Mortar Technology. Proceedings... CD ROM. Salvador: CEPED/EPUFBA/UCSAL/UEFS, 1997. p. 120-132.

KRUS, M.; HANSEN, K. K.; KUNZEL, H. M. **Porosity and liquid absorption of cement paste**. Materials and structures/Matériaux et constructions. Vol. 30. 1997. p. 394-398.

KUMMER, L.; BASSETTI, F. de J.; RIELLA, H. G.; AZEVEDO, J. C. R. **Reuse of porcelain and feldspar polishing residues in the manufacture of a new ceramic product**. Cerâmica Industrial. Magazine. Vol. 12. n. 3. May/June, 2007. Sao Paulo/SP. Available at: http://ceramicaindustrial.org.br/pdf/v12n3a05.pdf. Accessed on: June 28, 2008.

LANA, E. F. D. **Carbonation of the cover layer of concrete prototypes with fly ash and lime**. Santa Maria, 2005. 138 p. Master's degree dissertation. Postgraduate Program in Civil Engineering. Federal University of Santa Maria. Available at: http://coralx.ufsm.br/gepecon/DIS_Elinor2.pdf. Accessed on: June 27, 2008.

LANG, C. L.; ROCHA, J. C. **Characteristics of basalt crushed sand and dosage study for portland cement-based concrete for structural purposes**. Workshop: performance of construction systems. Chapecó/SC. November 27 and 28, 2006. Available at: HTTP://claudio.jacoski.googlepages.com/nove.pdf. Accessed on: June 27, 2008.

LEITE, M. B. **Evaluation of the mechanical properties of concrete produced with recycled aggregates from construction and demolition waste**. Porto Alegre, 2001. 270 p. Thesis (Doctorate). Postgraduate Program in Civil Engineering. Federal University of Rio Grande do Sul. Available at:

http://www.reciclagem.pcc.usp.br/ftp/Tese%20Monica%20Leite.pdf. Accessed on: July 2, 2008.

LEVY, S. M. **Contribuiçâo ao estudo da durabilidade de concretos, produzidos com residuos de concreto e masvenaria**. Sao Paulo, 2001. 194 p. Thesis (Doctorate). Department of Civil Construction Engineering. Polytechnic School of the University of Sao Paulo.

Available at: http://reciclagem.pcc.usp.br/FTP/tese%20SALOMON.pdf. Accessed on: July 1, 2008.

MARQUES, L. N.; MENEZES, R. R.; NEVES, G. A.; SANTANA, L. N. L.; LIRA, H. L.; FERREIRA, H. C. **Reuse of porcelain polishing residue for use in ceramic tiles**. Electronic journal of materials and processes. Vol. 22. 2007. p. 34-42. Available at:

http://www.dema.ufcg.edu.br/revista/index.php/REMAP/article/viewFile/39/69. Accessed on: July 9, 2008.

MEHTA, P. K.; MONTEIRO, P. J. M. **Concreto: estrutura, propriedades e materiais**. Sao Paulo: PINI, 1994.

MELO NETO, A. A. de. **Study of shrinkage in mortar with activated slag cement**. Sao Paulo, 2002. 161 p. Thesis (Doctorate). Department of

Civil Construction. Polytechnic School of the University of Sao Paulo. Available at: http://www.teses.usp.br/teses/disponiveis/3/3146/tde-05022003-114904/. Accessed on: June 27, 2008.

MENOSSI, R. T. **Use of basalt stone powder as a substitute for natural sand in concrete**. Ilha Solteira, 2004. 110 p. Dissertation (Master's Degree). Postgraduate Program in Civil Engineering. Faculty of Engineering of Ilha Solteira. Paulista State University. Available at: http://www.ppgec.feis.unesp.br/teses/2004/romulo_2004.pdf. Accessed on: June 19, 2008.

MIRANDA, L. F. R. **Study of the factors influencing the cracking of mortar coatings with recycled rubble**. Sao Paulo, 2000. 172 p. Master's degree dissertation.

Department of Civil Construction Engineering. Polytechnic School of the University of Sao Paulo. at:

http://www.reciclagem.pcc.usp.br/ftp/Dissertaçao_LEONARDO_miranda.pdf. Accessed on: July 1, 2008.

MOREIRA, J. M. S.; FREIRE, M. N.; HOLANDA, J. N. **Utilization of granite sawdust waste from the state of Espirito Santos in red ceramics**. Revista Ceràmica. Vol. 49. n. 312. 2003. p. 262-267. Available at:

http://www.scielo.br/pdf/ce/v49n312/a1249312.pdf. Accessed on: June 30, 2008.

MOURA, W. A. **Utilization of copper slag as an additive and as a mixed aggregate for concrete**. Porto Alegre, 2000. 232 p. Thesis (Doctorate). Postgraduate Program in Civil Engineering. Federal University of Rio Grande do Sul. Available at: http://www.reciclagem.pcc.usp.br/ftp/washington_ufrgs.pdf. Accessed on: July 2, 2008.

MOURA, W. A.; GONÇALVES, J. P.; LEITE, R. da SILVA. **Utilization of marble and granite cutting waste in coating mortars and making floor tiles**. Sitientibus. Feira de Santana. n.26. jan./jun. 2002. p. 49-61. Available at: http://www.uefs.br/sitientibus/pdf/26/utilizacao_do_residuo_de_corte_de_marmore.pdf.

Accessed on: June 30, 2008.

MÜLLER, A.; GLEIZE, P. J. P.; ROMAN, H. R. **Physical and mechanical properties of mortars with mineral and organic additions**. In: III Brazilian Symposium on Mortar Technology. Proceedings... CD ROM. Vitória: UEFES/PPGEC, 1999. Vol. 1. p. 309-322.

NAKAKURA, E. H.; CINCOTTO, M. A. **Analysis of the classification requirements for laying and coating mortars**. BT/PCC/359. Department of Civil Engineering, USP Polytechnic School. Sao Paulo, 2004. Available at: http://publicacoes.pcc.usp.br/PDF2003&2004_1/BT%20-%20359.pdf. Accessed on: July 3, 2008.

NASCIMENTO, M. R. do; GEMELLI, E.; CAMARGO, N. H. A. **Characterization of cement paste mortars with mixtures of bentonite and coal dust**. 51st Brazilian Ceramics Congress. Salvador, 2007. Available at: http://www.abceram.org.br/51cbc/artigos/51cbc-6-01.pdf .Accessed on: June 29, 2008.

NEVES, L.; BINA, P.; NOTRISPE, E. **Study of the use of various waste materials as components in concrete and mortar**. In: V Seminar on Sustainable Development and Recycling in Civil Construction. Proceedings... Sao Paulo: IBRACON/CT-206, 2002. p. 353-359.

NEVILLE, A. **Properties of concrete**. Sao Paulo: PINI. 1997.

OLIVEIRA, I. **Nature on the agenda**. FOCO magazine. 110. ed. Year VIII. Natal: FOCO. May 2008. Pg 26 and 27.

OLIVEIRA, M. L. L. **Influence of the addition of polypropylene fibers in mortars**. Florianópolis, 2001.   171 p. Master's degree dissertation. Postgraduate Program in

Civil Engineering. Federal University of Santa Catarina.

PAES, I. N. L.; ANDRADE, M. A. S. de; ANGELIM, R. R.; HASPARYK, N. P.; OLIVEIRA, R. A. de.; PASSOS, J. S.; THON, D. G.; CARASECK, H. **O efeito** de **finos calcârios nas propriedades da argamassa** de revestimento. In: III Brazilian Symposium on Mortar Technology. Proceedings... CD ROM. Vitôria: UFES/PPGEC, 1999. p. 335-347.

PANDOLFO, L. M.; MASUERO, A. B.; TIECHER, F.; PANDOLFO, A. **Properties of coating mortars produced with natural sand and basalt sand**. In: VI Brazilian Symposium on Mortar Technology. Proceedings... CD ROM. Florianôpolis: UFSC, 2005. p. 53-58.

PEREIRA, L. H.; JALALI, S.; AGUIAR, J. B. **Economic viability of a construction and demolition waste treatment plant.** December 2004. Available at: http://repositorium.sdum.uminho.pt/bitstream/1822/2596/1/Beja_BA.pdf. Accessed on: June 27, 2008.

PEREIRA, P. C.; IKEDA, J. A.; IKEDA, N. A.; CAMPOS, C. O.; TEXEIRA, L. M.; CARASEK,

H. **Cement content or a/c: who has** the **greater influence on bond strength?** In: III Brazilian Symposium on Mortar Technology. Proceedings... CD ROM. Vitória: UFES/PPGEC, 1999. Vol. 2. p. 489-502.

PETRY, S. B. **Study of water permeability in the cover layer of concrete prototypes with high fly ash contents**. Santa Maria, 2004. 111 p. Dissertation (Master's Degree). Postgraduate Program in Civil Engineering. Federal University of Santa Maria. Available at: http://coralx.ufsm.br/gepecon/Simone.pdf. Accessed on: June 27, 2008.

PINTO, J. A. N. **Elementos para dosagem de mortas**. 2. ed. Santa Maria. 1996. 143 p.

PINTO, T. de P. **Metodologia para a gestâo diferenciada** de **residuos sólidos da construção urbana**. Sao Paulo, 1999. 189 p. Thesis (Doctorate). Department of Civil Construction Engineering. Polytechnic School of the University of Sao Paulo. Available at: http://www.receiclagem.pcc.usp.br/ftp/tese_tarcisio.pdf. Accessed on: June 1, 2008.

PISSATO, E.; SOARES, L. **Addition of quarry fines in soil-cement mixtures**. In: IV Seminar on Sustainable Development and Recycling in Civil Construction - Recycled Materials and their Applications. Proceedings... Sao Paulo: IBRACON/CT-206, 2001. p. 331-339.

PONTES, I. F.; VIDAL, F. W. H. **Valorization of marble and granite sawmill waste and its application in civil construction**. Rio de Janeiro, 2005. Available at: http://www.cetem.gov.br/publicao/CTs/CT2005-101-00.pdf. Accessed on: June 27, 2008.

QUENARD, D. A.; XU, K.; KUNZEL, H. M.; BENTZ, D. P.; MARTYS, N. S. **Microstructure and transport properties of porous building materials**. Materials and structures/Matériaux et constructions. Vol. 31. n. 209. 1998. p. 317-324.

RESCHKE, J. S.; CECCATTO, D. M.; MASUERO, A. B.; MORAES, C. A. M.; VILELA, C. F. **Evaluation of the use of granulated foundry slag as a substitute for part of the cement in concrete**. e-Mat - Revista de Ciência e Tecnologia de Materiais de Construçâo Civil. Vol. 2. n. 1. May 2005. p. 78-91. Available at: http://www.antac.org.br/emat/e-MAT-V2-N1/e-MAT-V2-N1-p78-91(Res).pdf. Accessed on: June 27, 2008.

RIPPER, E. **Manual pràtico de materiais de construçâo: recebimento, transporte interno, estocagem, manuseio e aplicação**. Sao Paulo: PINI, 1995.

ROCHA, J. C.; JOHN, V. M. **Utilization of waste in civil construction**. Coletânea habitare. Vol. 4. Porto Alegre: ANTAC, 2003. 272 p. Available at:

http://habitare.org.br/pdf/publicacoes/arquivo/134.pdf. Accessed on: Nov. 14, 2008.

ROCHA, M. F. P.; OLIVEIRA, L. A. P. **Influence of the capillary absorption of ceramic blocks** on the **compressive strength of laying mortar**. In: III Brazilian Symposium on Mortar Technology. Proceedings... Vitória: UEFS/PPGEC, 1999. Vol. 2. p. 477-487.

RODRIGUEZ, C. O. **Los morteros, control de calidad**. First European cooperation. 1994.

ROMAN, H. **Masonry mortars**. In: III Symposium on the Performance of Building Materials and Components. Proceedings... Florianópolis: CTC- UFSC/DEC/NPC, 1991. p. 111-116.

ROMAN, H. R.; MUTTI, C. N.; ARAÙJO, H. N. de. **Building in structural masonry**. Florianópolis: UFSC, 1999. 83 p.

SA, M. das V. V. A. de. **Influence of replacing natural sand with stone dust on the mechanical, microstructural and electrochemical behavior of concrete**. Natal, 2006. 141 p. Thesis (Doctorate). Postgraduate Program in Materials Science and Engineering. Federal University of Rio Grande do Norte.

SABBATINI, F. H. **Laying mortars for resistant masonry walls**. IPT Bulletin, 1986, Sao Paulo, 26 p.

SAMPAIO, J. A.; ALMEIDA, S. L. M. de. **Obtaining artificial sand from quarry fines**. CETEM - Mineral Technology Center. Rio de Janeiro:2002. Available at: http://www.cetem.gov.br/publicacao/CTs/CT2002-194-00.pdf. Accessed on: July 9, 2008.

SANCHEZ, J. A.; BARRIOS, J.; BARRIOS, A.; AGUDO, A. R. A. **Shrinkage in lime mortars**. Construction Materials. Vol. 47. n. 245, January/February/March. 1997.

SCHENINI, P. C.; BAGNATI, A. M. Z.; CARDOSO, A. C. F. **Construction waste management**. Available at: http://geodesia.ufsc.br/Geodesia-online/arquivo/cobrac_2004/092.pdf. Accessed on: June 27, 2008.

SILVA, G. J. B. **Study of the behavior of portland cement concrete produced with the addition of porcelain polishing residue**. Belo Horizonte, 2005. 92 p. Master's degree dissertation. School of Engineering, Federal University of Minas Gerais. Available at: http://www.pos.demet.ufmg.br/hpcpgem/P/defesas/911M.PDF. Accessed on: July 9, 2008.

SILVA, N. G.; BUEST, G.; CAMPITELI, V. C. **Mortar with crushed sand: influence of fines and particle shape**. In: VI Brazilian Symposium on Mortar Technology. Proceedings... CD ROM. Florianópolis, 2005. p. 12-22.

SILVA, N. G. **Coating mortar made of cement, lime and crushed limestone sand**. Curitiba, 2006. 180 p. Master's degree dissertation. Postgraduate Program in Civil Construction. Federal University of Paranâ. Available at: http://dspace.c3sl.ufpr.br:8080/dspace/bitstream/1884/4660/1/DISSERTA%c3%830%20MES

TRADO.pdf. Accessed on: June 1, 2008.

SILVA, N. G; CAMPITELI, V. C. **Influence of fines and lime on the properties of mortars**. In: XI National Meeting on Technology in the Built Environment. Proceedings... CD ROM. Florianópolis, 2006. Pag. 4343 to 4358.

SILVA, N. G.; CAMPITELI, V. C.; GLEIZE, P. J. P. **Coating mortar made of cement, lime and sand from crushing limestone**. In: VII Brazilian Symposium on Mortar Technology. Proceedings... CD ROM. Recife: UFPE/ANTAC/GTARGAMASSAS, 2007.

SILVEIRA, R. G. da. **Study of the microstructure of the cover layer of concrete with high levels of fly ash**. Santa Maria, 2004. 164 p. Master's Degree. Postgraduate Program in Civil Engineering. Federal University of Santa Maria. Available at: http://coralx.ufsm.br/gepecon/Rodrigo.pdf. Accessed on: June 27, 2008.

SOUZA, P. A. B. F. **Study of the plastic, mechanical, microstructural and thermal behavior of concrete produced with porcelain residue**. Natal, 2007. 208 p. Thesis (Doctorate). Postgraduate Program in Materials Science and Engineering. Federal University of Rio Grande do Norte.

SOUZA, U. E. L.; FRANCO, L. S. **Subsidios para a optção entre: elevador ou grua, andaime fachadeiro ou balancim, mortar industrializada ou produzida em obra**. BT 176. Polytechnic School of the University of Sao Paulo. Sao Paulo, 1997. Available at: http://publicacoes.pcc.usp.br/PDF/btcap176.pdf. Accessed on: June 23, 2008.

SOUZA, V. C. M.; RIPPER, T. **Pathologia, recuperação e reforço de estruturas de concreto**. Sao Paulo: PINI, 1998.

TAYLOR, H. F. W. **Cement chemistry**. London: Academic press, 1990. 474 p.

TEIXEIRA, L. CBIC - Brazilian Chamber of the Construction Industry. **Analysis of the Cement Sector in Brazil**. Available at: http://www.cbicdados.com.br/files/textos/032.pdf. Accessed on: June 20, 2008.

THOMAZ, E. **Cracks in buildings**: causes, prevention and recovery. 1. ed. Sao Paulo: PINI, 2003. 194 p.

TRISTÃO, F. A. **Influence of the granulometric composition of sand on the properties of coating mortars**. Florianôpolis, 1995. 197 p. Dissertation (Master's Degree). Federal University of Santa Catarina.

TRISTÃO, F. A.; MACHADO, V. K. C. B. L. **Analysis of water retention and consistency methods in mortars**. In: V Simpôsio Brasileiro de Tecnologia das Argamassas. Proceedings... CD ROM. Sao Paulo: ANTAC, 2003. p. 233-242.

TRISTÂO, F. A.; ROMAN, H. R.; VALDAMERI, C.; MESACASA, C. **Use of industrial basalt sand in coating mortar**. In: VI Brazilian Symposium on Mortar Technology. Proceedings... CD ROM. Florianôpolis: ANTAC, 2005. p. 59-64.

VAGHETTI, M. A. O. **Effects of fly ash with rice husk ash or silica on the carbonation of Portland cement concrete**. Santa Maria, 1999. 118 p. Thesis (Doctorate). Postgraduate Program in Civil Engineering. Federal University of Santa Maria. Available at: http://coralx.ufsm.br/gepecon/marcos.pdf. Accessed on: June 27, 2008.

VOGT, J. C.; PINHO, V. X.; CORRÊA, E. C. S.; AGUILAR, M. T. P.; CETLIN, P. R. **Influence of the addition of mineral residues on the mechanical strength, fracture toughness and Young's modulus of concrete**. 17th CBECIMat - Brazilian Congress of Materials Science and Engineering. Foz do Iguaçu, 2006. p. 3756-3766. Available at: http://www.metallum.com.br/17cbecimat/resumos/17cbecimat-210-025.pdf. Accessed on: June 27, 2008.

YOSHIDA, A. T.; BARROS, M. M. S. B. **Characterization of mortars in the fresh state: peculiarities in the analysis of industrialized mortars**. In: I Simpòsio Brasileiro de Tecnologia das Argamassas. Proceedings... CD ROM. Goiânia: ANTAC, 1995. p. 53-62.

ANNEX A

Essay	Method	Unit	Results	Specification NBR 11578 (ABNT, 1991)
Loss on ignition - PF	NBR 5743 (1989)	%	5,24	$\leq 6,5$
Magnesium oxide - MgO	NBR 9203 (1985)	%	1,33	$\leq 6,5$
Sulfuric anhydride - SO_3	NBR 5745 (1989)	%	3,27	$\leq 4,0$
Insoluble residue - IR	NBR 8347 (1992)	%	0,94	$\leq 2,5$
Alkaline equivalent in Na O_2	-	%	1,04	not applicable
Free calcium oxide - CaO	NBR 7227 (1990)	%	1,34	not applicable

Source: Report provided by the manufacturer, March 2008.

Table A1 - Chemical characterization of Portland cement.

Essay	Method	Unit	Results	Specification NBR 11578 (ABNT, 1991)
Specified area (Blaine)	NBR 7224 (1996)	cm^2/g	4026	≥ 2600
Fineness - retained on the 0.075 mm sieve (#200)	NBR 11579 (1991)	%	1,3	$\leq 12,0$
Fineness - retained on the 0.044 mm sieve (#325)	NBR 11579 (1991)	%	7,7	not applicable
Normal consistency paste water	NBR 11580 (1991)	%	28,4	not applicable
Start of grip	NBR 11581 (1991)	h:min	3:06	≥ 1
End of grip	NBR 11581 (1991)	h:min	4:11	≤ 10
Lê Chatelier expandability - hot	NBR 11582 (1991)	mm	0,17	$\leq 5,0$

Source: Report provided by the manufacturer, March 2008.

Table A2 - Physical and mechanical characterization of Portland cement.

Age (days)	Minima	Màxima	Average	Deviation	Specification NBR - 5736 (ABNT, 1991)
3	28,7	33,3	31,7	1,00	$\geq 10,0$
7	35,4	39,1	37,8	1,09	$\geq 20,0$
28	43,0	47,3	45,5	1,07	$\geq 32,0$

Source: Report provided by the manufacturer, March 2008.

Table A3 - Portland cement compressive strength.

ANNEX B

Essay	Results
Fineness - residue on the 0.044 mm sieve (#325) (%)	1,50
Humidity (%)	2,00

Source: Report provided by the manufacturer, March 2008.

Table B1 - Physical characterization of CH I lime.

Composition	Results
RI - insoluble in hydrochloric acid (%)	0,5
Magnesium oxide (MgO) (%)	3,5
Silicon dioxide (SiO_2) (%)	0,3
PPP (%)	23,0 - 25,0
$Ca(OH)_{2d}$ (%)	90,0 - 98,5
R2O3 (%)	1,5

Source: Report provided by the manufacturer, March 2008.

Table B2 - Chemical characterization of CH I lime.

ANNEX C

Classification of mortars according to NBR 13281 (ABNT, 2005).

Class	Apparent mass density in the 3rd hardened state kg/m	Test method
M1	≤ 1200	
M2	1000 a 1400	
M3	1200 a 1600	NBR 13280
M4	1400 a 1800	
M5	1600 a 2000	
M6	> 1800	

Table C1- Apparent mass density in the hardened state.

Class	Tensile strength in bending (MPa)	Test method
R1	≤ 1,5	
R2	1,0 a 2,0	
R3	1,5 a 2,7	NBR 13279
R4	2,0 a 3,5	
R5	2,7 a 4,5	
R6	> 3,5	

Table C2 - Tensile strength in bending.

Class	Mass density in the state fresh kg/m^3	Test method
D1	≤ 1400	
D2	1200 a 1600	
D3	1400 a 1800	NBR 13278
D4	1600 a 2000	
D5	1800 a 2200	
D6	> 2000	

Table C3- Mass density in the fresh state.

Class	Water retention %	Test method
U1	≤ 78	
U2	72 a 85	
U3	80 a 90	NBR 13277
U4	86 a 94	
U5	91 a 97	
U6	95 a 100	

Table C4- Water retention.

Buy your books fast and straightforward online - at one of world's fastest growing online book stores! Environmentally sound due to Print-on-Demand technologies.

Buy your books online at
www.morebooks.shop

Kaufen Sie Ihre Bücher schnell und unkompliziert online – auf einer der am schnellsten wachsenden Buchhandelsplattformen weltweit! Dank Print-On-Demand umwelt- und ressourcenschonend produzi ert.

Bücher schneller online kaufen
www.morebooks.shop

Printed by Books on Demand GmbH, Norderstedt / Germany